YOUR DAI

ALSO BY

*BEND YOUR BRAIN*

# 24 HOURS IN THE LIFE OF YOUR BRAIN

THE MINDS BEHIND marbles the brain store®
WITH GARTH SUNDEM

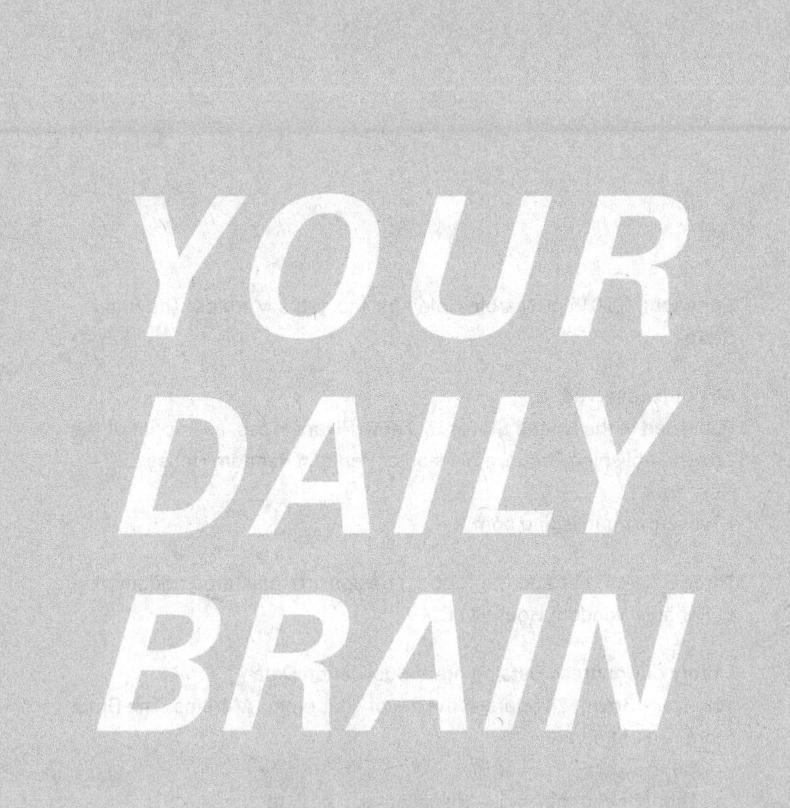

THREE RIVERS PRESS • NEW YORK

Published in the United States by Three Rivers Press, an imprint of the Crown Publishing Group, a division of Penguin Random House LLC, New York.
www.crownpublishing.com



Library of Congress Cataloging-in-Publication Data
Your Daily Brain : 24 hours in the life of your brain / Marbles: The Brain Store. —First edition.
pages cm
1. Thought and thinking—Miscellanea. 2. Attention—Miscellanea. 3. Intellect—Miscellanea. 4. Mental health—Miscellanea. 5. Brain—Miscellanea. I. Marbles: The Brain Store.
BF441.T724 2015
153.4—dc23 2015004795

ISBN 978-0-8041-4011-9
eBook ISBN 978-0-8041-4012-6

Printed in the United States of America

Cover design by Marbles: The Brain Store

10 9 8 7 6 5 4 3 2 1
First Edition

# CONTENTS

# INTRODUCTION

The marimba has always been such a soothing sound—cloth mallets against wood bars, perfect for making the sounds of raindrops in forest pools. Then the iPhone went and co-opted it for its default alarm. Now the sound of a marimba makes an entire train full of commuters check their jacket pockets.

Consider this book your marimba. It's time to wake up, jump-start your brain, and rip back the veils of habit, misperception, and irrationality that your neurons create between you and reality. But waking up is just the start. From the second the marimba sounds until your eyes close for the evening (and beyond!), you evaluate the world outside, evaluate the world inside, make choices and decisions, plan how to reach little goals like getting to work on time, and keep your eyes on bigger goals like being a good parent or partner, all amid a landscape of distraction and temptation. Every second of every day, you have the opportunity to use your brain for better or for worse. You have the opportunity to nail it or totally screw up. And when you nail these things or screw them up, you have the opportunity to learn.

The thing is, someone somewhere has studied each one of these teeny-tiny chunks in the day of your brain. From multitasking over your morning coffee to milking the semiconscious state before you sleep for flashes of insight, there's someone with a PhD in a white lab coat—or, more likely, jeans and a T-shirt—who knows how to do it better. This book collects science's best understandings of how to maximize the use of your brain, generally organized by the situations in your day when you're likely to use these skills. Some of the entries are fun facts that you can use to dazzle in conversation over a platter of crudités, some offer understanding or a new way to look at the things you do and why you do them, and some suggest little actions or exercises that can help you use your brain to manage the challenges of your day and your life better.

Here's *Your Daily Brain,* a fun way to shine light into those dark corners of your mind usually penetrated only by cranial nerves, and a chance to explore how our evolving understanding of the brain can help you live a better life.

*PART 1*

# THE MORNING

## *WAKING UP, GETTING READY, GETTING GONE*

## *INSIGHT OR ENERGY: SHOULD YOU HIT THE SNOOZE BUTTON?*

You've heard of brain waves, and here's how they happen. Think of the eighty-six billion neurons in your head like crickets. When one cricket chirps, not much happens—it's not like the voice of one cricket can make you pour a cup of coffee or help you remember that snappy comeback. But your brain crickets don't just do their own thing. They synchronize in ways that create the pulsing cacophony of a summer night.

The thing is, crickets can come into synch in different rhythms. The neuronal crickets in your brain "chirp" more slowly when you are asleep than they do when you are awake. Just like a summer night, the brain waves created by your chirping neurons are like the background noise against which other things take place. When you're awake, everything you do or think happens against the backdrop of the pattern called beta waves. Deep sleep happens against delta waves. If you listened closely, the beta waves would sound like high-pitched chirps and the delta waves would sound like crickets bowing a section of orchestral basses. Between these two patterns—the beta waves of alertness and the delta waves of deep sleep—are alpha waves of wakeful relaxation and theta waves of light sleep.

So there are many patterns of brain waves created by the synchronicity of your neuronal crickets, and each brain wave is associated with a level of sleep or consciousness. The purpose of an alarm is to mess with these crickets, forcing them to chirp in the pattern you want. Of course, you have a last line of defense against the dictatorship of your alarm clock: the snooze button! The desire to whack snooze competes only with the need to check Facebook while driving and with the overwhelming compulsion to scratch mosquito bites on your knuckles for the top spot on the human list of temptations. The question is, should you?

Here's the reason you shouldn't: maybe if you set the alarm five, ten, or fifteen minutes later, you wouldn't need the alarm at all. If you kicked the snooze habit, you could sleep a little longer, and these few minutes might be all it takes for your brain to reach a natural state of wakefulness without being tossed into the ice-water bath that is the alarm clock. If you stick to a regular sleep schedule, your body knows exactly when it's reached the final pass through what's called N1 sleep, and you'll wake up instead of taking another spin through the sleep cycle. If the fifteen minutes that you usually spend hitting the snooze button would let you get into this final N1, your brain and body would be better off using this time to sleep for real instead of snoozing in the half-light.

Then again, there's something to be said for hanging out at this N1 transition between alpha and theta waves. Have you ever been floored by an idea? Has insight ever hit you like a falling piano? When did it happen? Was it in a warm shower or in the middle of the night? The reason is that a brain coasting on the cushion of theta waves is primed for insight—when you relax in the shower or slip into the boundary between alpha and theta waves in N1, you make your brain ready to receive messages from the beyond. There you are between sleep cycles in an N1 phase or staring out the window cross-eyed at a rainy day, and *wham!*—it's insight (which looks like a burst of high-frequency gamma waves in your brain).

If you need energy, forget the snooze button and work toward a regular sleep cycle that lets you wake up naturally. But if you need insight, try smacking snooze and surfing the cusp between theta waves and alpha waves—the line that separates asleep from awake. You may find your brain infusing the certainty of insight into what had been stubbornly murky before.

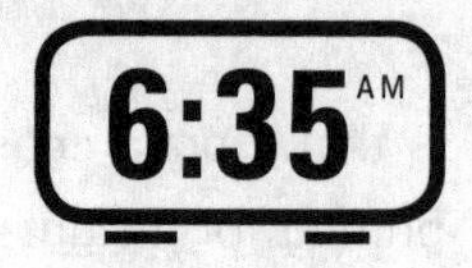


## *THE DIFFERENCE BETWEEN ASLEEP AND AWAKE*

Why are we here? Are we alone in the universe? What is consciousness? Why do cat pictures go viral on Facebook? A satisfactory answer exists for exactly one of these questions—namely, the consciousness thing. And it explains what happens when you come back into your body in the morning.

The understanding comes from a long line of research by Francis Crick (of discoverer-of-DNA fame) and Christof Koch at the Allen Institute for Brain Science in Seattle. They started with a simple question: Is there one area of the brain that lights up during all the tasks of consciousness? If all these sensory and motor and cognitive things were circles of a Venn diagram, where would they overlap?

What they found is the claustrum, a one-millimeter-thick sheet of neurons that divides the hemispheres of the brain. All mammals have it. And it's connected to all the major players in your skull, including the prefrontal cortex, auditory cortex, visual cortex, primary motor cortex, premotor cortex, and many other areas of functioning. So Crick and Koch went into people's heads and zapped their claustrums to see if it would mess with their consciousness. Actually, they didn't—due to troublesome things like ethics and morality, you can't just fry people's brains and see what happens. That is, outside of very special circumstances.

One of those circumstances is in the treatment of

epilepsy. In epilepsy, an area of the brain misfires in a way that lets electricity "leak" into surrounding tissues, rebounding through the brain like a gunslinger's bullet in a rib cage—sometimes with equally devastating consequences. To treat cases of severe and debilitating epilepsy, doctors explore inside the brains of conscious patients to discover the source of the problem, at which point they can sometimes cure or diminish symptoms by inserting a sophisticated electrical pacemaker. The thing is, epilepsy can live pretty much anywhere in the brain, and so discovering its source sometimes takes significant exploring.

That's what Mohamad Koubeissi and colleagues did with a fifty-four-year-old with what they describe as "intractable epilepsy." In a study published in *Epilepsy and Behavior* in 2014, they recount what happened during what's called electrical stimulation mapping, when the surgeons just happened to be poking around her claustrum. "Stimulation of the claustral electrode reproducibly resulted in a complete arrest of volitional behavior, unresponsiveness, and amnesia without negative motor symptoms or mere aphasia," they write. In English, this means that when Koubeissi zapped this woman's claustrum, she became unconscious. When they turned off the juice, she was again immediately conscious. During brain mapping, patients usually read aloud or do some other kind of brain task that can show doctors how their prodding affects function. In this case, as Koubeissi introduced high-frequency electrical impulses into his patient's claustrum (i.e., "frying"), she would stop reading and stop responding to her surgical team, and her body would gently slow into a state of deep relaxation. When the signals stopped, she would open her eyes and be able to continue reading.

In a *Forbes* article, Koubeissi called the claustrum the "sleep switch" and likened it to turning the key in a car's ignition. The understanding of this difference between asleep and awake, conscious and unconscious, is still in its early stages, and opportunities to double-check the finding don't come around so often. But here's a cool part: now that we are beginning to understand the location and function of human consciousness, it might make it possible to not only understand the roots of our own consciousness but also replicate this consciousness. Knowing the difference between asleep and awake in your brain may make it possible for us to artificially create "awakeness."

## Tongue Slips

**Outside the big switch of Consciousness with a capital *C* are those little slips of consciousness, like when you call your knowledgeable friend "a vast *suppository* of information." Or when you turn *spaghetti* into *pasketti*. The first is called a malapropism (when you switch around meaning), and the second is a metathesis (when you switch around sounds). Listen for these today.**

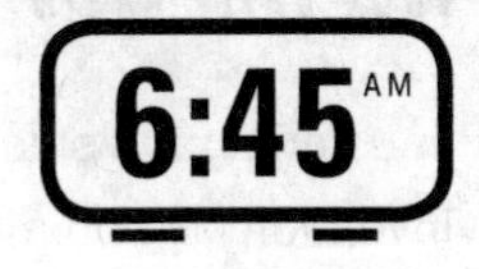


## USE THE POWER OF NEWNESS TO WAKE UP YOUR SENSES

When you wake up, your brain is bored. And just because your eyes are open in the morning doesn't mean there's anything going on behind them. In fact, studies have demonstrated that even when researchers can show that areas in the primary visual cortex are taking things in, your brain may remain unaware. Your brain can have the physical input of "seeing" without being conscious of seeing.

The same is true of hearing, smelling, tasting, and touching: once you get used to something, you stop recognizing it. If you live near the airport, you may stop hearing airplanes. If you live near the paper factory, you might stop smelling heated pulp. And a big part of why scratching an itch makes the itch even worse is the awareness that scratching causes. Without the directed action of rubbing your fingernail across the itch, you get used to the itchy feeling, which eventually fades into the unnoticed background of your brain. Scratching the itch makes it new and makes you notice it again.

This means that in the morning, you are combating not only a brain that may have been harassed from sleep by the brutality of an alarm clock but also the humdrum habituation of eight hours of sameness in your bedroom. So if you want to wake up, give your senses something new. This may be as simple as thinking a new thought—just as you can't go to sleep at night if thoughts of your next great invention or book idea are crashing around in your head, bringing interesting, engaging, creative thoughts into your brain first thing in the morning can help you exorcise the grogginess of an empty brain. Consider priming your brain pump with an interesting idea in a notebook left open on a bedside table.

Or you can force your brain to deal with something new in the world. A skateboard at the top of the stairs should do it. For something a bit less extreme, setting your alarm to an unpredictable radio station or investing in a system that clobbers your olfactory bulb with a new scent upon waking can challenge your brain to get online and start making sense of things.

Finally, don't disregard the power of anticipation. If your brain is more motivated by what it is being forced to leave behind (namely, your warm bed) than it is by the desire for what it anticipates ahead (namely, a frenetic morning and a day filled with work . . .), it may resist waking up. Instead of mourning the loss of your bed, find something to look forward to, even if it's just coffee and lots of it.

## Wake Up with . . . Stress!

If a new thought, new sensory experience, or something to look forward to doesn't wake you up in the morning, try stress. A study of British civil service employees showed that stress-induced cortisol levels in the brain made people able to get up early for work. But it's a dark magic: too much stress-induced cortisol may degrade the brain over time.

### *DO YOU CROW WITH THE ROOSTERS OR HOOT WITH THE OWLS?*

You know, I know, we all know that the whole "morning person" vs. "night owl" thing is all in the brains of people who happen to like late-night talk shows or love those ninety minutes in the morning before the kids get up. Science agrees: it *is* all in their brains. But that doesn't mean it ain't real. For example, folks who consider themselves morning people literally create the chemical experience of "awakeness" faster than people who consider themselves night owls. It's all about the chemical cortisol—morning people have higher saliva cortisol levels than night owls an hour after waking up. This is true even when night owls and morning people get exactly the same amount of sleep and report the same quality of sleep.

And check this out: according to the National Center for Education Statistics, the average start time for U.S. public high schools is 7:59 a.m. You've probably heard that this start time is cruel to teenagers, and if you look in teens' brains, you can see why. Just alongside the area of the teen brain that stores incomprehensible text message abbreviations and the area that keeps hyperactive track of how to look hot is a perfect storm of factors that makes early rising not only inconvenient but diabolically cruel for people suffering from being between the ages of twelve and eighteen.

We are all at the mercy of circadian rhythm: no matter

how much we have slept or haven't slept, our brain wants to sleep when it gets dark and wake up when it gets light. There are two ways we can see circadian rhythm making you sleepy: an increase in the hormone melatonin and a decrease in body temperature. Put the two together and you have a darn precise picture of when the body wants to go to sleep.

And melatonin goes up and body temperate goes down in teenagers' bodies later than it does in the bodies of normal human beings. That's the first gust of the perfect storm. The second is that teenagers need more sleep than older or younger humans, about eight and a half to nine and a half hours per night, according to the American Academy of Pediatrics (AAP). If a teenager who needs more sleep is physiologically blocked from getting to sleep until 11:00 p.m. and then has to get up at 7:00 a.m. in order to make the 8:00 a.m. bell, there's no way outside time travel to get enough sleep. The teenage brain is already wired for impulsivity. Now add chronic sleep deprivation and, again according to the AAP, you've got a recipe for depression, obesity, low test scores, and, basically, a scenario straight out of one of the post-apocalyptic novels the kids are all reading these days.

If it weren't for the mechanics of the adult workday that make it so much *easier* to have kids in school at 8:00 a.m., and the need for teenagers to be at home watching television by 4:00 p.m. at the latest, any rational and compassionate person would start the school day later—say, 9:00 a.m. at the earliest.

The thing is, there may be a teenager hiding out in your brain no matter what age you are. In other words, it's unfair to say that *all* adult-ish people can get up early while it's best for *all* teenagers to stay up late. These are just the broad brushstrokes—your experience may vary. And if you simply can't seem to wake up in the morning, it very well may be that your brain wants you to be a night owl.

That said, there are some good things about being a morning person. Studies show that morning people are generally more stable, responsible, agreeable, and optimistic, but then there's that old chicken-and-egg question: does being a morning person make you agreeable, or do agreeable people happen to be morning people? This means that if you're wired to be a night owl, maybe waking up early in hopes of making yourself optimistic and agreeable is only going to make you an unhappy, sleep-deprived night owl. If you can't bend your brain to your life, ask if you can bend your life to your brain. If you're a night owl, can you admit your needs and sleep in a little? Try telling your employer that inside your head lives the brain of a teenager. When that doesn't work, ask how you can streamline your morning so that you can let your night owl self stay in bed just a little longer.

## Melatonin for Jet Lag?

Your circadian rhythm adjusts to when you eat. A Harvard study showed that you can reset your circadian rhythm by fasting for twelve to sixteen hours. Then when you wake up and eat, your body jumps back online as if whatever time you wake up is the natural morning. You can also hack your melatonin level. A study in the *British Medical Journal* found that taking a melatonin supplement for three days before traveling at exactly the time you will be going to sleep at your new destination can help you adjust to jet lag.

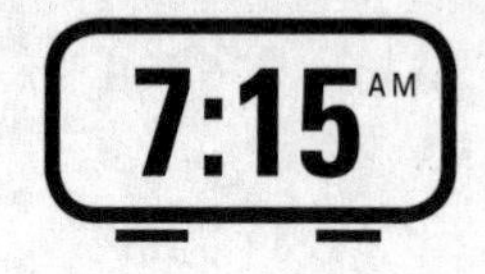


### *HERE'S WHAT YOUR BRAIN LIKES FOR BREAKFAST*

Your brain runs on pure, high-octane sugar, and lots of it. Although your brain accounts for only about 2 percent of your body's mass, even at rest it consumes about 20 percent of the body's energy. Glucose passes through the blood-brain barrier into your brain cells and then into the little independent organisms called mitochondria that live in your cells, where it is turned into the basic cellular energy called ATP in a process called the Krebs cycle, which you may remember reading on your forehead in the mirror after sleeping on your high school biology textbook. In any case, your brain can't really store glucose and so, like a hummingbird, it needs a steady drip. When you eat sugar your brain gets a glucose flare and then it peters out, like one of those Estes model rocket engines. And if you don't eat at all, well, your body can metabolize some of its fat resources to keep your brain from turning into a turnip, but it ain't exactly ideal.

All this is to say that your brain needs a good breakfast. From book smarts to decision making to processing speed to memory to willpower, if you don't feed your brain, it will punish you. Study after study shows that the unfed brain underperforms the fed brain on just about any task you can imagine. If you want to get nitpicky, brains that are fed breakfasts with a low glycemic index outperform brains that are fed only sugary foods. For example, a study of 290 children published in the journal *PLoS ONE* shows that kids who ate low-glycemic-index, rice-based breakfasts had higher gray matter volumes and even higher overall IQs than kids who ate high-glycemic-index, primarily bread-based diets. And when UK researchers compared Cheerios to a glucose drink to no breakfast in twenty-nine schoolchildren, they found that the glucose-drink and no-breakfast groups crashed and burned on cognitive tests performed later in the morning.

***BRAIN BREAKFAST***

**Yes:** Old-fashioned oatmeal, whole-wheat toast, raisin bran, grapefruit

**No:** Instant oatmeal, bagel, sugary cereal, apple juice, raisins

*Source:* Harvard Medical School glycemic index for 100+ foods

It's not that a little sugar is a great and horrible brain killer, to be avoided like mercury in shellfish. In fact, it's been shown that a *little* sugar functions as a nice a.m. pick-me-up, and that may be why cultures around the world have evolved cuisines that tend to include a bit more sweetness in breakfast than in lunch or dinner (think syrup-drenched pancakes and two lumps in the coffee). But it's a bit like the old idiom about hats and cattle: all hat (sugar) and no cattle (low-glycemic-index foods) and your morning wakefulness may be short-lived. In terms of breakfast, if you're going to wear the hat, you'd better also have the low-glycemic-index cattle to go with it.

## Eat Breakfast, Lose Weight

At least skipping breakfast helps you lose weight, right? Wrong. Even when you pull out all the socioeconomic and demographic factors you can imagine might be different between people who eat breakfast and those who don't, people who don't eat breakfast are across the board more likely to have a high BMI. There have even been FDA-approved clinical trials that show *prescribing* breakfast helps promote weight loss. How can eating help you lose weight? Simply, eating breakfast makes your brain able to resist unhealthy snacking later in the day. If you want to be thin, start by being realistic: eat breakfast.

### *MORNING! PANIC! MULTITASK!*

Make breakfast. Pack lunches. Feed the pets. Groom the offspring. Groom the spouse. Groom self. Check email. Solve world hunger. Drink more coffee. Ponder. Get dressed. Remember you need half-and-half. Schedule the afternoon.

Morning is go time. And it's never, ever time to go in any single direction. Instead, the morning is the time to multitask. Some of us are better at it than others. . . .

In fact, there really is a medical term to describe those of us who can't multitask. It's called *strategy application disorder*, and it's a kind of frontal lobe dysfunction in which people show "disorganization, absentmindedness and problems with planning and decision making in everyday life despite normal performance on traditional neuropsychological tests," according to research published by Paul Burgess in the journal *Psychological Research* in 2000. Take a second to digest that. The lovely person in your life who is normal but for the fact that he or she can't possibly do more than one thing at a time may have a brain that is shaped differently than yours. Your loved one may be biologically programmed to fold laundry, make sure the guinea pig has been brought in from the outside run, or think about the weather, but not any of these things simultaneously. You wouldn't make fun of someone with a concussion, and you shouldn't make fun of your monotasking loved one either. If you do, it makes you a bad person.

On the opposite end of this spectrum is a group of people recently identified as *supertaskers*. According to a 2014 study by University of Utah psychologist David Strayer, just as the monotasking brain is different from most brains, the supertasking brain is too—exactly 2 percent of us have areas of the anterior cingulate cortex and posterior frontopolar prefrontal cortex that make us able to do many things at once. The rest of us 98 percent may have exactly the same brainpower as supertaskers in terms of IQ and other goodies, but when we add task A to task B, we take longer and perform worse than if we had focused on task A and then transitioned completely to task B.

The trick to successful multitasking is not, as you might expect, splitting your mind into two sections, each in charge of monitoring and completing a demanding task. The trick is switching the brain's attention between tasks so quickly that it *looks like* you're doing two things at once. You start with your spotlight on task A and swivel the light to task B, moving back and forth so quickly that it appears you can spotlight both at the same time. Psychologists call this *task switching*.

When you switch from one demanding task to another, it takes time to get oriented—you pay a time and/or an accuracy penalty, and both tasks suffer. That is, unless you're a supertasker. There's debate whether you can actually change your brain enough to go from being an average-tasker to a supertasker, but there are certainly rules that can help you fake it. Here are four things you can do to improve your multitasking ability:

1. If you can make cognitively demanding tasks less cognitively demanding, you'll pay a lower penalty when you combine them. It makes sense: if packing lunches is automatic, you'll be able to do it while cooking an omelet. If you have to think about lunches, it takes away from your ability to think about the omelet. By making any demanding task more automatic, you reduce its demands on your attention and so have more attention for a second (and third, and fourth . . .) task.

2. Pick tasks that use faraway areas of the brain—say, a motor task like flipping the omelet along with a cognitive task like remembering which kid likes which snacks. If two tasks are far enough apart in your brain, it's almost like you're not even doing two things at once.

3. Practice "fixing" tasks in your mind. For example, if you can remember exactly where you left off making an omelet, you can learn to come back to exactly that point without paying the "switching penalty" that would otherwise be involved in reorienting yourself. Practice saying "I am chopping onions" or "When I come back I will add cheese" before switching to another task, and it should make it easier to switch back to the omelet later.

4. Outsource the monitoring that tells you *when* to switch tasks. If you're worrying about when you need to flip the omelet while packing lunches, packing lunches is going to suffer. The trick is called *task cuing*—set your

kitchen timer to remind you when the omelet is ready to flip (cue when you should switch tasks). If you are helping to bring your strategy-application-disordered loved one up to speed, task cuing might look like an ongoing narrative that reminds this person what he or she should be doing at any given second. Try it: your loved one will totally thank you.

### Multitask Training

If you're willing to do a little work to improve your multitasking ability, search online for the phrase "task switching" to find games like the Trail Making Test, the Wisconsin Card Sorting Task, and the Stroop Color and Word Test. Basically, these games force you to switch nimbly between sets of rules while inhibiting distractions. Task-switching practice makes for nimble multitasking.

## *THE (REVERSE) POWER OF POSITIVE THINKING*

It's morning and the possibilities of the day stretch before you. What do you want to get done today? Or at the very least, what do you want to work toward? Is it things like losing weight, getting a job, recovering from injury, or getting a date? In each case, studies show that the more you imagine positive outcomes, the *less likely* you are to achieve them.

"Positive thinking fools our minds into perceiving that we've already attained our goal, slackening our readiness to pursue it," writes NYU psychology professor Gabriele Oettingen in an article for the *New York Times*. For example, when she asked students to write about what a happy week would look like, and then tested them at the end of the week, it turns out they ended up feeling less energized and got less done than students who had written about a realistic week.

The same is true of those positive affirmations recommended by Stuart Smiley on *Saturday Night Live*, which studies show have an effect exactly opposite of what they're intended to do. Apparently when you look in the mirror and tell yourself that you're good enough and people like you, it can activate your understanding that you may not actually be good enough and that people don't like you. Your brain knows when you're blowing smoke up your own backside. And when positive affirmations are too far removed from the reality in which your brain knows that you live, positive self-talk is more depressing than bolstering.

But then on the opposite side of this seemingly dismal science of positive thinking is optimism. Optimistic people score higher on tests, have better mental health overall, and live longer. The difference between positive thinking and optimism, as Oettingen shows, is that one depends on *fantasy*, while the other depends on *expectation.* For example, in Oettingen's famous study of obese women trying to lose weight, she measured their expectation of eventually reaching their goal weight and also their fantasies of magically reaching their goal weight. In this case the word *expectation* is a lot like *optimism*—predicting good things for the future—and women in Oettingen's study who expected good things achieved them. But women who fantasized about waking up to find they were thin were less likely to have lost weight a year later.

So ask yourself about your goals this morning. What do you expect and what do you fantasize about? If your mind's wanderings are too far removed from reality, then those fantasies may be as close as you get to the real thing. But if you can ground your optimism in an outcome that your brain actually expects you to achieve, it can act like the carrot pulling you forward toward your goal.

## I Got This, I Got This . . .

Positive self-talk isn't without its place. Studies of dart throwing and endurance exercise show that chanting versions of the "I got this, I'm going to kill it" mantra to yourself increases performance. Just make sure you limit positive self-talk to actions and not self-perceptions. Self-talk can force you to visualize the best outcome and improve actions. Self-perception isn't so easily tricked.

### *HOW TO PLAN YOUR DAY*

The morning is ground zero for planning. What will your second grader bring for sharing? When will you pick up the dry cleaning? Do you have time to work out today, and if so, should you pack your good running shoes or your second-string gym shoes? When will the dogs get a walk so they don't go marauding for stuffed things to tear apart and spread around the backyard while you're at work? Who will pick up the kids, or should they take sleeping bags and toothbrushes to school so that you can pick them up tomorrow? The list is endless, and it's a miracle of the brain that some of the plans you make in the morning may even come to pass as you design them.

The skill of successful planning depends on something called *executive function*, which lives in a little chunk of your frontal lobe. In the morning, you use your executive function to look into the future—you imagine the future you desire, generate options for behaviors that could create this future, and finally decide on the best option to get from where you are now to that idyllic future you imagine (in which everyone is happy, fed, cleaned, educated, exercised, enriched, and in bed by 8:30 p.m.).

Without executive function, you can't look into the future, you can't plan, and so you become impulsive, acting in ways that pay off right now but may be bad in the long term. In short, without executive function, you become a teenager: impulsive and incapable of making decisions that will seem good in hindsight. Teenagers have an excuse, since the prefrontal cortex, which holds executive function and its skill of planning ahead, isn't fully developed until your mid-twenties. Technically, the teenage brain is still "pruning" pathways to make executive function more efficient. And this means that for better or for worse, as you set out on the great journey that is your teenage years, all the possibilities of executive function still stretch out before you. It still seems like maybe it's a good idea to ride a skateboard off the roof while a cute girl is watching. Or maybe it's fine to ditch your friend to dance with that really hot guy at a concert. The teenage brain hasn't yet pruned away these possibilities. It hasn't hardwired the ability to look into the future to predict consequences, and so it can't plan—it can't visualize the future it wants, generate strategies to get there, and pick the one that's most likely to lead from A to B. First it needs to get rid of all those synapses that lead down side roads and around roundabouts, those pesky inefficient synapses that make teenagers the stars of the vast majority of "fail" videos on YouTube.

This morning as you're planning the mechanics of your day, your only excuse for failure is forgetting to use your prefrontal cortex. The key to successful planning is to put your future self in charge—to imagine in the morning that you're looking back at the day with hindsight. What will you wish that you'd done? For example, studies of grocery buying show that intentionally imagining that your future self in charge makes people buy healthier food. The other lesson from studies of planning is that the only way to truly improve is with practice: only planning really makes you better at planning. Start by being conscious of what works and what does not. Can you really combine picking up the dry cleaning with buying dog food on the trip home? Can you really walk the dogs enough in that middle-of-the-day break to keep them from eating the house? Like a teenager's brain, the more you prune away the inefficient neural pathways that leave your kids stranded at soccer practice and your dogs unwalked, the less *fail* you will find in your ability to successfully plan the day.

### Train Your Body, Train Your Brain

A 2011 study published in the journal *Science* collects all the strategies proven to boost executive function. It includes things you might expect, like computer-based brain training. But it also includes things like aerobic exercise and mindfulness training. It seems that an efficient body translates to an efficient mind, at least when it comes to the executive-function skill of planning.

## *THE PROBLEM WITH PUZZLES*

Do you wake up with Sudoku over your morning coffee? Do you do the *USA Today* or *New York Times* crossword on the train? Congratulations! You've probably gotten very good at Sudoku or the crossword! Now, if your daily life depends on being able to put numbers or letters into grids, you are tons better off for all this practice. Otherwise, not so much. Your practice with these puzzles might boost your confidence in your brain skills, and like a sugar pill, any placebo that you believe works is very likely to work. But other than the placebo effect, your puzzling prowess remains little more than a parlor trick.

The reason is something called *transfer*—do the brain skills needed to complete a puzzle transfer to the skills of your life? And how far does the brain training of a puzzle transfer? Unfortunately, your brain is pretty darn bad at transferring the lessons from any skill you practice into any other skill, no matter how closely related. For example, do you remember the video game Tetris? It's that game where blocks fall from the sky and you have to stack them so that they fit together in complete rows. Believe it or not, Tetris has been studied pretty extensively as a test of visual-spatial skills, specifically (as you might imagine) as a test of figure rotation, or your mind's ability to imagine what something will look like turned on its side or flipped on its head.

Let's take a quick look inside the brain training of Tetris. What we see is pretty representative of any type of puzzle. First, there's the hopeful angle: many studies show changes in the brain after Tetris training, most notably decreased activation in frontal and parietal areas of the brain that are associated with mental rotation tasks. In other words, after playing Tetris, you get more efficient at rotating these shapes in your mind, and it takes less brainpower to imagine what they look like when they're spun. That's so cool! All it takes is twelve hours of Tetris training to see real, physical changes in the brain.

And this Tetris training makes subjects better at parallel parking, packing luggage into an airplane's overhead bins, figuring out how frozen food fits in the freezer, and organizing Tupperware containers of holiday decorations in the garage. Actually, it doesn't. Not at all. Tetris training is so specific to Tetris that it doesn't even make you better at rotating shapes that aren't seen in the game. And when people trained in Tetris rotate these shapes in their minds, they spin them clockwise, just like in the game.

The moral of the story is that Tetris makes you better at Tetris. And Sudoku makes you better at Sudoku. Even the crossword puzzle makes you better at the crossword (though in the case of the crossword, there's some evidence that it makes you better at the *process* of recall, so chalk this one up in the maybe column). The lesson is that any puzzle you do makes you better only at that puzzle.

Alas! Is there no hope for biggering your brain this morning over your morning cup of joe? Can you not puzzle your way to happier, more numerous neurons? Actually, you can, and the trick is both simple and intuitive: while practicing a puzzle makes you better only at the puzzle, learning how to do *a new kind* of puzzle challenges your brain to constantly make sense of new information—and if there's anything you should take away from this book, it's that constantly doing new things is good for your brain. Once you understand how to do Sudoku, it's time to leave Sudoku behind. Once you understand Tetris, it's time to move on. It's the process of figuring out *how* to do a new thing and not the rote practice of things you already understand that will make your brain more facile with the kinds of things that really matter to your life.

## This Morning: A Puzzling Puzzle

This morning, over your cup of coffee or during the morning train commute, search out a kind of puzzle you've never tried before. Maybe try to finally understand what is going on with the north-south-east-and-west bridge puzzle in the morning paper. Or search online for "bridge puzzle" to find a neat new kind of puzzle, popular in Japan, that you might not have seen before. Then give it an hour or a day or a week—long enough that you understand the instructions and feel comfortable with some of the nuances. At that point, you've harvested the puzzle's brain-training ability and it's time to move on to another type!

## *HOW TO REMEMBER WHERE YOU PUT YOUR CAR KEYS*

You spent all that time planning, and now as you prepare to launch from the house this morning, all your plans have puffed into the gentle breeze of forgetting like the seeds blowing off a dry dandelion. Gone, gone, gone is the memory of that 9:30 a.m. phone call and the fact that your child is supposed to bring a white T-shirt to use in art class today. Gone even is the location of your car keys, though you vaguely remember that you set them somewhere odd. Could they be on the bathroom shelf? No, that's not quite right. Maybe on the laundry room folding table? Almost. This section will help you hold on to your morning plans, even minutes or hours after you make them! Consider it a brief morning memory refresher.

Of course, the topic of memory is a huge one, and there are dozens of brain books written just on the subject of memory training alone. In part that's because "memory" is actually a handful of related but not identical skills. The way you remember a smell is different from the way you remember how to shoot a free throw, which is different from the way you remember the list of your day's activities.

But all types of memory share the three general stages of encoding, storage, and retrieval—how you put it in, hold on to it, and get it out—and for each of these stages there are ways you can learn to do it better. The key is hacking the brain structures involved in each task.

For example, your hippocampus is the structure that shoves facts into your head. Your senses take in experiences and then your hippocampus packages these experiences for storage deeper in the brain. The problem is, your senses bombard your hippocampus with more than it can possibly handle, and so it has to sort experiences into ones it will help you remember and ones you can immediately forget. The hippocampus does this by "tagging" things with little priority stickers. You can help your hippocampus tag the things you want to remember. For example, if you set your car keys in the refrigerator's butter dish, you may never find them again, but simply saying to yourself "I am setting my car keys in the butter dish" forces the hippocampus to tag the memory with a higher priority. You are more likely to remember the things you do with consciousness than the things you do automatically or without really meaning to.

But if you want to store memories with absolute top priority, the tags of consciousness aren't enough. For that, you need the tags of emotion. To get them, you'll need the help of another brain structure—namely, the amygdala. Some people call the amygdala your "lizard brain," because the two marble-shaped balls of the amygdalae are the home of fear and desire, the emotional learning centers of your brain. If you were trapped in an elevator when you were nine, it's your amygdala that makes you claustrophobic now (it's also the emotional response of the amygdala, balanced by rational thoughts from your prefrontal cortex that creates many of our decisions, as we'll see later). And if you spent a horrified hour looking for the car keys that you accidentally threw out with the morning trash when late for an important meeting, it's this emotional memory that can help you remember where you put them now. If you really want to remember anything, make it emotional.

**MEMORY TIPS**

+1 Tagging

+1 Emotion

–1 Stress

People who win memory competitions know this trick. To memorize the order of a shuffled pack of cards, memory champions tell themselves lewd and emotionally charged stories in which the cards are characters. You may not remember queen of diamonds, three of clubs, jack of spades, but you're not likely to forget the storyline of Madonna showing her breasts to George Clooney (that is, if your brain includes Madonna's "Material Girl" to represent the queen of diamonds, the image of a busty three of clubs, and George Clooney as a charming rascal like the jack of spades). It makes sense that emotional memories are lasting: remembering the fear you felt when you nearly fell from the upper branches of a tree can keep you from doing it again.

The only exception to the rule of emotion-is-memory is stress. If a memory is attached to a *single* stressful event, like the search for your keys, the event will be indelibly inked in your brain. But against a general backdrop of chronic stress, you don't learn as well—it's as if this stress takes over part of your ability to create new memories. So stress is a complex dance: specific stress paired with specific memory creates things like PTSD flashbacks, but a high general baseline of stress makes the events that take place in this soup harder to remember.

We've talked about encoding and storage, and now it's time for recall—your ability to pull memories from storage deep in your brain. Long-term memory is retrieved by association. Something makes you think of something else, as in feeding your dog his meds smushed in a blob of butter and then suddenly remembering that you put your keys in the butter dish. Cued retrieval is also why the song "Stairway to Heaven" makes you think of middle school dances and why going to a baseball game could make an alcoholic fall off the wagon.

So to remember where you put your car keys this morning, you can try chilling out so that your memory isn't plugged by stress, and you can try attaching emotions to these memories. Then, when these strategies don't work, you give in to the realities of the modern brain and make a list. That's cued retrieval: a couple of chicken scratches on paper can cue the information you need to remember. Or do the same thing with string tied around your finger or a small rock in your pocket. Now where did you put that list again?

## Make a Spatial Memory

fMRI tests of memory champions show that it's spatial skills that let them work their magic, as if memory champions are walking through a complex house full of rooms rather than looking at a deck of cards. The house-full-of-rooms isn't far from the actual strategy: by constructing a story with characters and movement, memory champions take abstract symbols and make them meaningful. For example, after competing in the 2006 World Memory Championships in the Examination Schools of Oxford University, twenty-three-year-old Ed Cooke told *Discover Magazine* how he does it: Cooke assigns each card a person, action, and thing—for example, the queen of clubs is his friend Henrietta, the action is whacking with handbags, and the thing is a closet full of designer clothes. He memorizes cards in groups of three—the first card is represented by the person, the second by the action, and the third by the image. So he might say, "Destiny's Child is whacking Franz Schubert with handbags," as in the *Discover* article. When trying to memorize abstract symbols or things, give them meaning—make these things represent emotionally charged memories and you will remember them.

### *SETTING YOUR POSITIVITY CLOCK*

We don't measure ourselves in absolutes. From wealth to happiness to success to the feeling of fulfillment at the end of a long day, we measure ourselves against our expectations. It's like the stock market: it may be not a company's earnings that drive its stock price but how these earnings compare to forecasts. You can see it in customer service too. A classic 1993 article in the *Journal of Marketing Research* notes that a customer's perception of customer service is "a function of a customer's prior expectations of what will and what should transpire during a service encounter." If customers *expect* that their call will probably be dropped, they will be happy just to reach someone. If customers *expect* that the service representative will come to their house to clean the bathroom floor, even a no-questions-asked product return policy won't be good enough.

And let's look at people who lose seven pounds on a weight-loss program. A 1997 study showed that if people were told they would lose five pounds, they were overjoyed to lose seven, and if they were told they would lose ten pounds, they were disappointed. Finally, in terms of "subjective well-being"—how people report feeling about their lot in life—it all depends on what they expect, so much so that people in China have historically reported the same subjective well-being as people in the United States. This is despite a median household income in the United States of $84,300 and in China of $10,220 (2009 numbers from *Forbes*), and an average 12 years of schooling in the United States and only 6.4 years in China. In the United States, we just expect more—and we're dissatisfied when we don't get it. Due to expectations, average life satisfaction is consistently measured as being higher in India than it is in Japan.

Okay, okay, just one more example. Imagine your favorite sports team loses a home game. How do you feel about it? Well, UC San Diego economist Gordon Dahl shows that it matters whether your team was favored or not. What was the *expected* outcome? If the home team is an underdog and then loses, it's no big deal. But when a favored home team loses, Dahl showed there is a spike in the city's rate of domestic violence. When a sports team doesn't live up to its expectations, fans get mad in a visceral and ugly way.

What this means is that this morning and every morning, you have the opportunity to set expectations for your day that will affect whether you feel good or bad about it this evening. But it's tricky—you don't want to lower your expectations just so that you can be pleasantly surprised. Lowering your expectations may result in lowered results. For example, expecting to earn a median household income of $84,300 and complete twelve years of school may help us actually achieve these things. It may be that expecting good customer service means that you will receive good customer service. And if you expect a C minus, you may be chuffed with your C plus, but it still ain't gonna get you into Harvard.

So you're walking a tightrope here, with alligators on both sides of the river below. Expect less and you may get less. But if you expect too much, you're setting yourself up for disappointment.

The solution is realism. If you set realistic expectations and then fail to meet them, maybe the disappointment will motivate you to do better next time. And when you exceed these expectations, you can still be pleasantly surprised. Will you eat only one unhealthy snack today? Will you run three miles without kidding yourself that you'll run ten? Will you spend fifteen minutes talking on the phone with your mother? Being realistic lets you succeed as much as you fail, earning the pleasure of overperforming while keeping the motivation of sometimes missing your goals.

## Expect Process, Not Product

So what happens when you set realistic expectations and then bad luck makes you fall short? I mean, that happens, right? And there's no need to be disappointed when a bad result really, truly isn't your fault. One solution is to set your realistic expectations for *process* and not *product.* If you expect to try your best, you can still be pleased when bad luck makes you fail. Or if you expect to work efficiently, you can still go to bed successful even if a couple of tasks remain in your in-box. This goes for your expectations of others as well. For example, studies show that if you praise a child's effort and not innate qualities like intelligence, the child will be more likely to work hard toward goals instead of depending on (questionable) smartness to carry the day. At the end of the day, recognizing and rewarding these elements of process will help you succeed more often when it comes to product.

## *WHY YOU CAN'T REMEMBER STREET NAMES OR DIRECTIONS*

You're out the door! You did it! Now if you could only remember where you're going . . . But if you can depend on your smartphone to tell you to "turn left in six feet" when walking from your kitchen to the restroom, is your ability to remember driving directions superfluous? Let's all hope the answer is yes, 'cause we're sure as heck losing this ability. Studies show that, due in large part to what are called "egocentric" maps—ones that know where you are—the human brain is losing its ability to navigate by landmarks and compass directions.

This ability depends on something called "place" and "grid" cells. In 2014, the discovery of these cells earned their discoverers the Nobel Prize in Medicine or Physiology. In 1971 John O'Keefe discovered the brain cells that tell us where we are, and in 2005 the husband-and-wife team of May-Britt and Edvard Moser discovered the cells that help us get where we're going. When you drive to work or school you use both these kinds of cells—or at least you used to. Here's how they work. O'Keefe's "place cells" are nerves in the hippocampus that are active when you are in a specific position—your hippocampus holds little maps of the places you know, and each place on those maps corresponds with a different little cluster of hippocampal nerve cells that fire together. When you go to a new place, you construct a new mental map and your brain assigns different clusters of nerve cells to represent these places in your hippocampus. Theoretically, though not yet practically, we should be able to look inside a person's hippocampus and, from the pattern of nerve activation, discover where that person is located.

The Mosers picked up where O'Keefe left off and started their research with a simple question: what is connected to the hippocampus and its place cells? To find the answer, they looked inside the brains of rats as they navigated mazes. Near the hippocampus is something called the entorhinal cortex, which you can remember because it sounds like an operation that would be performed inside a large, single-horned African odd-toed ungulate. It's been known that among the functions of the entorhinal cortex (EC) are things like monitoring the direction in which your head is pointing and helping to form conscious memories of the places you've been; it's an important connection between your hippocampus and your prefrontal cortex, the home of your consciousness.

When the Mosers watched the ECs of maze-running rats, they saw something very cool: as rats would pass one marker and then the next and the next, certain cells within the EC would flash. These cells, it turned out, represented points on a hexagonal grid "similar to the hexagonal arrangement of holes in a beehive," notes the Nobel Foundation. By lighting up in certain patterns, the grid could keep track of how an animal was moving and could allow animals to plan the steps they should take to get from one place cell "here" to another place cell "there." The Mosers called these "grid cells."

We can see a version of these grid cells in the brains of humans playing video games. Using microfine electrodes, researchers are able to record the firings of single neurons. Just like in rats running a maze, when researchers look at the spittings of a single neuron in the entorhinal cortexes of humans playing video games, they can tell, for example, if the human is making a character run clockwise or counterclockwise.

So if you want to get from your house to work, place cells will tell you where you are and grid cells will tell you how you are traveling.

Interestingly, one way to ensure that your brain's GPS remains accurate is to get enough sleep. Imagine that a rat activates a sequence of place cells while running a maze. When sleeping, the rat's brain will activate these same cells in the same order as it consolidates this spatial memory. The next day, rats that have been allowed to sleep navigate mazes they remember more successfully than rats that haven't slept.

Another key to navigating without your smartphone is to *navigate without your smartphone.* Again, it's about the hippocampus. Studies show that London cabbies who are forced to memorize reams of spatial information as part of the licensing test known as The Knowledge have a higher than average volume of gray matter in their hippocampi. And people who navigate by the stimulus-response strategy of their phone's GPS have smaller hippocampi than people who call up a mental map while navigating. Not only does texting while driving make you more likely to smash into things (more on that later), but letting your smartphone tell you how to get where you're going may mean that your brain becomes less able to find where you're going without it.

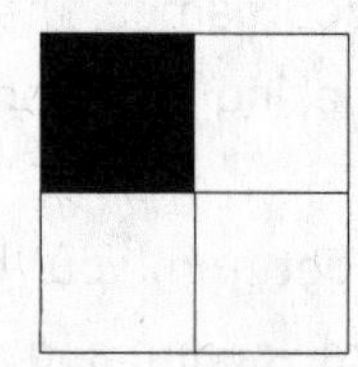

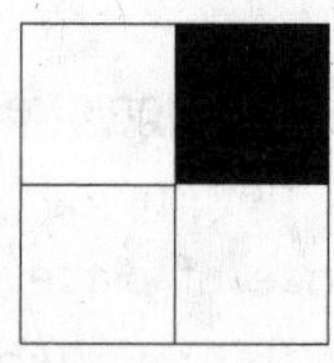

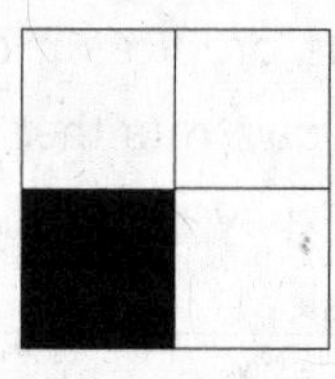

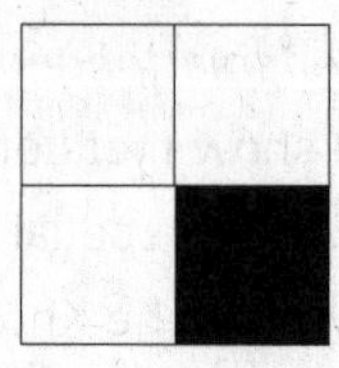

## Visual Pattern Span Test

Unlike London cabbies, you don't have to remember the spatial layout of "every possible route through the city as well as memorizing landmarks and points of interest, museums, parks, police stations, churches, theatres and schools and not just the famous landmarks like Buckingham Palace and Nelson's Column" (from theknowledge taxi.co.uk). But it's nice to be able to get from home to the coffee shop without depending on your phone's blue dot. If you've temporarily misplaced this ability (i.e., you've lost your phone), try some training with the Visual Patterns Test. It's like Simon Says, but you'll have to make your own patterns. Copy and cut out the boxes on page 56 (or make your own on graph paper), and then place these boxes in long strings of patterns. By tapping upper right, upper left, lower right, and lower left, how many boxes in a row can you reproduce before you forget what comes next?

### *HOW TO PICK A CAR RADIO STATION*

What do you want during your morning commute? Do you want to be soothed or enriched? First, the solid stuff: people with certain personalities prefer certain kinds of music. For example, a study of 1,443 people from twelve to nineteen years old found that people who prefer "elite" music, including jazz and classical, tend to be agreeable, conscientious, and open to new experiences, but less emotionally stable. Those who like rock music, while open to new experiences, are much less conscientious. Teenagers who preferred urban or pop/dance music were extroverted and agreeable. During the three years of this study, the few kids who measurably changed their personalities also changed their music preferences.

One big question is whether change flows the other way: by changing the music you listen to, can you change your personality? Part of the puzzle is that we like what we have heard before. For example, a famous study published in the journal *Memory and Cognition* asked people to listen to melodies. Then after some time had passed, the study played back these melodies along with distractor melodies composed to sound similar but not identical to the target melodies. Subjects were asked to rate how much they liked each melody—and sure enough, even if they didn't consciously know which they'd heard and which were new, they liked the melodies they'd heard before. This is why you like music from your youth and teenage years: you've heard it before, it's familiar, and familiarity breeds enjoyment.

So forcing yourself to listen to Mozart every morning on the way to work will eventually make you like Mozart. But the question remains: will it eventually make you the kind of person who likes Mozart? Over the long term, we really don't know. What we do know is that in the short term, music can have a huge effect on your mood. Studies show that listening to familiar music is soothing and that listening to new music can evoke anxiety. Here's another thing we know about the music in your car: listening to the music you know and enjoy for the average American commute of 25.4 minutes is almost half an hour in which your brain isn't really forced to do anything. Those sing-along tunes you know and love may help you regulate your mood (especially because you may attach these songs to personal, emotional memories), but they don't do much to grow your brain. New music forces you to expand your mind. Just like learning a new language, listening to new music can push your brain out of its prewired comfort zone, spurring it to grow new connections between neurons that may not have spoken to each other since you decided that you are a Pearl Jam person and not a Kenny Rogers person. If you want to be stroked in a soothing way, listen to the music you know well. If you want to grow your brain, try a new station.

## Musical Mind Control

Today, try to listen to the soundtrack that plays in the background of your life. And as you do, recognize that every choice of music has a purpose: it's meant to make you do something. Is the music in your taxi meant to please you or the driver? What does that company mean to convey with its on-hold music? Do you find yourself humming along with music at the coffee shop? And what stations or genres do you gravitate toward at different points in your day? This stuff isn't chance. For example, fast music can make you happy, but it increases your perception of time spent on hold; about seventy to a hundred beats per minute seems to be about ideal for on-hold music. And a host of studies show the effects of music on spending—in restaurants, listening to classical music makes diners more likely to order "sophisticated" entrees, which, of course, cost more. Again, rather than letting the ambient music of the spaces you inhabit wash over you, today try to *listen*. By becoming conscious of this ambient noise, you can choose to let it influence you or to firewall yourself against the feelings and actions this music is meant to create.

## *WHY AMERICAN BRAINS COMMUTE ALONE*

*Get your motor runnin' . . . head out on the highway! Looking for adventure . . . and whatever comes my wa-ay!* C'mon, sing along: you know the Steppenwolf classic. When you hear it, you can't help but picture the dotted yellow line stretching out to the horizon. In contrast, there are far fewer freedom-evoking lyrics written about, say, taking the BART.

But if you really want to know why Americans are averse to public transportation, the best thing to do is ask them. That's what a 2007 study in the journal *Transportation Research* did—it asked nineteen single-car commuters to opine in an open-ended way about why they chose to drive, and then it looked at the transcripts of these conversations to see what themes of drivership emerged. The five reasons most cited for car commuting were "journey time concerns; journey-based affect [driving is emotionally easier]; effort minimization; personal space concerns; and monetary costs," the article notes. However, the article goes on to point out that if time and money are concerns, then the commuters in this situation would have been much better off taking the train.

And "better off" doesn't stop at the mechanical stuff. A 2010 Gallup poll shows that the longer a person's commute, the greater the back pain, fatigue, worry, and likelihood of being obese. In another study, the famed psychologist Daniel Kahneman had workers track their positive and negative emotions and showed that the ratio was most skewed toward the bad during the commute than at any other point during the day. And the MIT economist Robert Frank showed that accepting a longer commute in exchange for the higher square footage you could afford in a house in the suburbs is a bad trade-off when it comes to well-being: it's easy to project your expectation of enjoyment and appreciation onto a large house far away from your work in the city, but once people have made the decision, it turns out that accepting a longer commute for the privilege of a larger house reduces well-being.

The effects of the single-car commute have been measured in almost every way you can imagine (including the fact that we're gassing ourselves), and in almost every one of these measurements, commuting *totally sucks*. Why oh why would we be so dead set against exploring alternatives? There's no rational answer. But there are a host of *irrational* answers, and in the brain these tend to do just fine, thank you very much.

Here's one: we know what it's like to do the single-car commute, and we don't know what it's like to take the train. Economists have looked at a twist on this: when we're choosing a larger house farther away versus a smaller house closer, we know the exact difference between the square footage of the two houses, but we can only guess at the difference in well-being that will be created by the longer commute. Your brain is *risk averse*, meaning that there is a little thumb on its scale of rationality to make you hold on to what you have rather than risking it for something you *could* have. Would you bet $100 on a coin flip to win $201? If not, you're risk averse. And this risk aversion makes you cling to a house's knowable square footage while underweighting the unknown effect of the extra commute time.

The second reason we stay stuck in our one-car commute is that we don't actually measure our commute by how awful it is—we measure it compared to how awful the commute was for everyone else. A 2011 paper in *Transportation Research* shows that overall commute satisfaction depends on comparative happiness. When you look inside the study, there's an especially interesting point: car commuters are happiest when comparing their commute times to those of other car commuters, some of whom had to drive farther. And commuters who use non-motorized forms of transport or take public transportation are happiest when comparing themselves to car commuters. This means that a major reason we stay stuck in car commuting is because everyone else stays stuck in car commuting, and at least you don't have it as bad as Frank from accounting. The ability to compare your commute to a herd of other car commuters makes your commute look not so bad. The *Transportation Research* study also implies that if you were comparing your commute to people who took the train, your overall commute satisfaction would be lower and maybe you would eventually do something about it.

Gandhi told us to *be the change you wish to see in the world*. Bill Murray told us to *be the ball*. Somewhere in this mix of initiative and imagination is the courage to explore options other than the single-car commute.

## The Swedish Commute

Compare the United States to Sweden. In the United States, as we've seen, emotions are more negative during the commute than at any other point during the day—a long commute is like taking a chain saw to well-being. In Sweden, a study in *Social Indicators Research* of the country's three largest urban areas shows that "satisfaction with the work commute contributes to overall happiness." Blame it on the car. As the study's author writes, "Possible explanatory factors include desirable physical exercise from walking and biking, as well as that short commutes provide a buffer between the work and private spheres." The single-car commute may be your only option. But if it's not, maybe it's time to explore the alternative—that is, if you want a boost in happiness and well-being.

### *SHOULD YOU MAKE OUT WHILE DRIVING?*

Albert Einstein once said, "Any man who can drive safely while kissing a pretty girl is simply not giving the kiss the attention it deserves." Or maybe it was Mark Twain. Or Yogi Berra. Or Lao Tzu. In any case, who said it is beside the point. The real point is this: you have a limited pool of attention, and any attention that you use for activities other than driving is attention that you take away from driving.

That makes the question of distracted driving seem pretty cut-and-dried, but like most questions about the brain, there's more to it. Whether or not you should make out while driving depends on how much attention it takes to drive, how much attention it takes to kiss, and how much attention you have to go around. If the attention needed to drive plus the attention needed to osculate is less than the pool of attention you have to spend, why not smooch behind the wheel?

Of course, what we're really talking about here is texting. Can your brain safely text while driving? And how does texting compare with making out? Here's a short answer from the *American Journal of Public Health*: "After declining from 1999 to 2005, fatalities from distracted driving increased 28 percent after 2005, rising from 4,572 fatalities to 5,870 in 2008. Crashes increasingly involved male drivers driving alone in collisions with roadside obstructions in urban areas. We predicted that increasing texting volumes resulted in more than 16,000 additional road fatalities from 2001 to 2007."

Them's pretty strong words. And if you are a male driver prone to texting as you drive alone in urban areas, maybe you should stop right here. But the thing about statistics is that while they're pretty good at describing the view from ten thousand feet of what tends to happen in large populations of people, they're pretty awful at making predictions about what will happen with any individual person. Overall, it's not a good idea for people to text while driving. But what about *you*? Are you on a special place on the bell curves of attention, skill, and driving, so that texting while driving is bad for other people but fine for you?

How much are you willing to bet?

One thing that can cloud your judgment is overconfidence in your skills. Did you know that about 80 percent of people are "above-average" drivers? Of course that's impossible. But studies show that a disproportionate percentage of people *believe* they are above-average drivers. Which is to say that you may not be fully rational when it comes to the evaluation of your own skills behind the wheel of a motor vehicle.

Another thing that affects your bet is just how much a smartphone throws off your mojo. An article in the journal *Environment and Behavior* shows that even the presence of a smartphone on a table reduces the quality of a conversation taking place. If even something as important as trying to hook up over a dinner date is crushed by the mere presence of a cell phone, do you really think you are immune to the distraction of texting while driving?

When you put these two facts together you get a situation in which you can't really trust your evaluation of your driving skill and you can't really trust your opinion that of course your smartphone isn't a distraction. And the truth is you really don't *know* if, as you drive alone in urban areas, you are as much at risk as a male driver prone to texting or if you are somehow excused from the statistics. The brain's inability to be rational about its own abilities and challenges behind the wheel means that even when you don't feel like you're risking it, you might be risking it. Now, making out is another question altogether, and until solid research comes out against it, there seems no need to contraindicate something so *nice*.

## Fear or Rules for Stopping Teen Texting?

Want to make your teenager stop texting while driving? An article in the *Academy of Marketing Studies Journal* shows that your best strategy depends on whether the teenager is a boy or a girl. The study of 840 young adults shows that girls were most receptive to "fear appeals" via social media—for example, showing pictures of terrible texting-while-driving accidents—whereas young men were "much more likely than females to suggest using laws and legal action to discourage distracted driving." Rules or fear: match the right appeal to the right teenage gender and you've got a better chance at stopping texting while driving.

## *HOW TO PARALLEL PARK*

There's no way to avoid it: now is the time to talk about stereotypes. And two in particular—namely, the notions that men have better spatial abilities than women and that men are better at parallel parking than women. Let's unpack these stereotypes and see if it's phrenology or fact. First, the argument for fact. All sorts of tests show that men consistently score higher than women on tests of spatial ability (one example among many is a much-cited 2003 study in *Memory and Cognition* by two Canadian researchers, both of whom happen to be female). If you ask whether most men have better spatial visualization skills than most women, the answer, independent of your ideology, opinion, and gender-informed perspective, is yes. Unfortunately, this difference isn't just an interesting observation—with and without being studied in tandem with gender, spatial ability has been shown to predict high school geometry skills and scores on mathematics sections of college entrance exams. Spatial skills have real-world consequences even beyond scratching the paint on your car's bumper, and when you line up all men against all women, most men do it slightly better.

The question, then, is *why*. Is there something innate or genetic in the male brain that makes it better able to imagine the possibilities of shapes? We can certainly see mental rotation tasks in the brain—when researchers from the Montreal Neurological Institute watched people's brains with fMRI imaging as they matched figures to their rotated forms, blood flow in the brain increased in the right postero-superior parietal cortex, the left inferior parietal cortex, the left inferior parietal region, and the right head of the caudate nucleus. You won't necessarily be tested on those names, but suffice it to say we know where the brain processes spatial information. And while the results are a bit inconclusive, there's not really much difference between the brains of men and women as they rotate shapes in their minds. And there's certainly no male gene linked with spatial ability.

So if gender difference in spatial ability doesn't necessarily live in the structural capabilities of our brains, where does it live? One answer is that instead of nature, the difference may be due to nurture. Maybe the way men tend to grow up in our culture trains them in spatial abilities in a way that women miss. While many questions involving the brain and culture are met with a fuzzy soup of competing findings, we actually know the answer to this, and it comes from a 1,300-person study published in the *Proceedings of the National Academy of Sciences*—in other words, it's pretty darn solid.

In 2010, researchers from the University of California at San Diego traveled to northeast India to compare people from the Khasi and Karbi tribes. The tribes live next door to each other in the hills surrounding Shillong, and genetic tests show that they are closely related. In the Karbi tribe, women are not supposed to own land and the eldest son inherits property—they are a patrilineal society. In the Khasi tribe, property is inherited by the youngest daughter, men are not allowed to own land, and men are supposed to hand over their earnings to their wife or sister—they are a matrilineal society. This kind of natural division makes researchers drool. And after drooling (or perhaps concurrent with drooling), the researchers had 1,279 people from these villages solve a visual puzzle—anybody who could do it in less than thirty seconds earned a bonus of twenty rupees, about a quarter of a day's wage.

Here's what they found: in the patrilineal Karbi tribe, men had better spatial visualization skills, whereas in the matrilineal Khasi tribe, the spatial skills of women were just as good as the skills of men. Furthermore, there were a couple of households in the Karbi tribe that bucked the patrilineal norm. In these few Karbi households in which women happened to own land and money, the gender gap in spatial skills was only a third the size it was in the more common, male-dominated households.

The gender gap in spatial visualization skills is nurture, not necessarily nature. And even in our highly evolved Western society, we remain more Karbi than Khasi. In fact, the gender gap in spatial abilities goes beyond training—there's an unfortunate difference in how the sexes perform on tests as well. See, on some level many of us are aware of this male-dominated stereotype, and both men and women bring this awareness with them to, for example, college entrance exams. When we bring preconceived notions of skill to a pressure-filled situation, something unfortunate happens; it's called *stereotype threat*. Your brain has only so much brainpower to go around, especially in the area of your working memory, which is the part of your consciousness where you manipulate information in a goal-directed way. You need the full power of your working memory to do well on tests, and your working memory famously has limited slots—these are the "chunks," like digits or other bits of information, that you can keep in the front of your mind at any time. Research shows that worry literally claims space in your working memory, pushing out other useful information or processing ability. This means that the more worried you are that you will do poorly on a test, the more likely you are to do poorly on the test.

Now replace the word *worry* with the words *stereotype threat*. In many cases of stereotype, it has been shown that people who fit a negatively stereotyped group worry that they will perform according to the negative stereotype, thus blotting out space in their working memory and helping to ensure that they perform according to the negative stereotype. In this case, women who say to themselves "I'm bad at geometry" are worried when they hit the geometry section of the SAT or ACT, and their worried working memories have less space to find the correct answers.

Does this mean that women are doomed to underperform men in spatial abilities, score lower on math tests, and remain underrepresented in fields such as engineering, science, and technology? Hardly. It's perfectly possible to blast training and stereotype threat out of the water, for example by doing something as simple as playing a video game for a couple of hours. Researchers from the University of Toronto tested men and women on a spatial visualization test and found that men scored better than women. Then they had both groups play an action video game that required zooming around a three-dimensional space. After ten hours of gaming, the gender difference in spatial visualization skills was eliminated.

Yes, this example is a bit glib—we're not going to solve the gender gap in STEM fields by playing video games. But it shows that the gender gap in spatial abilities is real and that it can be erased by offering men and women exactly the same training.

Think about this in terms of parallel parking. What training is offered to men and women? To answer this, imagine what happens in most male-female relationships in the United States when both parties get in a car. Do you think that in *most* relationships the man drives or the woman drives? That's the training part. And now imagine the testing part—are you aware of the stereotype that men are better at parallel parking than women? If you are, then you help to create a self-fulfilling prophecy. Or you can decide there's enough repression due to violence inherent in the system, say the heck with parallel parking, and get a car that will do it for you.

## The Paper Folding Test

Search online for "paper folding test"—there's a great version at spatiallearning.org. In this ingenious and devious task, a piece of paper is folded and then a hole is punched in the folded paper. What will the punched paper look like when unfolded? Matching the folds and punch to the correct target forces you to bend your mind's spatial skills along with the paper. Do it enough and you will improve your spatial abilities. Whether or not this translates to better parallel parking may depend in large part on what you believe. If training your spatial skills makes you feel like a better parallel parker, it can smush the influences of stress and stereotype threat in your brain, making you a better parallel parker.

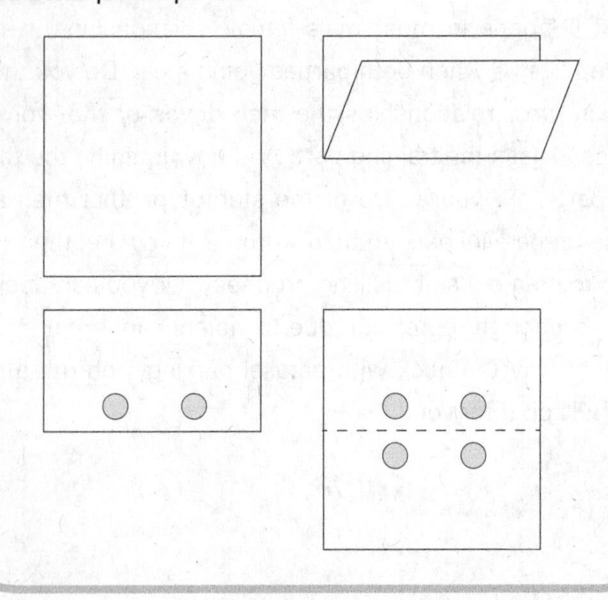

PART 2

# THE DAY

## *WORK, THINKING SKILLS, AND WILLPOWER*

## *HOW TO PAY ATTENTION*

As any child knows, even the sunniest day isn't truly scorching; to fry an ant, you need a magnifying glass. Something similar is true in your brain—you can have all the IQ in the world, but you have to focus its power to get things done.

Like many skills described in this book, the challenge of attention is choosing how to distribute a limited resource across many possibilities. Think of it like spending your holiday gift-buying budget on family and friends. Unless you are Bill Gates or you have a very limited gift list, you probably can't spend $100 on every person. Instead you spend more on some and less on others. You have to make the same difficult choice with attention. For example, imagine you're in a meeting. The more attention you spend on any one thing, the less you have to spend elsewhere. What do you do? Well, first you distribute your attention fairly evenly, spreading it across as much of the scene as you can take in at once. There are things that might grab your attention, like a coworker spilling her coffee, but barring an obvious assault on your attention, you apply your personal search criteria to the scene. If you're attending the meeting, you might notice who's there. If you're running the meeting, you might notice who's *not* there. And when you come across something that fits these criteria, you zoom in. Now your attention is more focused—a smaller slice of the scene consumes a larger part of your attention, and you use the same brainpower to get more information from less space.

If you are especially interested in getting information from this thing that is personally relevant to you, your focus might be so narrow that you're blind to other things you might like to notice in the environment, such as the coworker's coffee that is tipping into your laptop. This is called *inattentional blindness*, as seen in the famous "invisible gorilla" experiment, in which subjects who are asked to watch a video of people passing a basketball and count the number of passes fail to notice a person in a gorilla suit walking through the game.

Inattentional blindness goes even further. In a 2012 study published in the journal *Psychological Science*, Harvard researcher Tafton Drew asked twenty-four radiologists to read films and look for lung nodules—a familiar task. On the last film, Drew inserted the image of a gorilla forty-eight times the size of the average nodule. Eye tracking showed that the vast majority of these twenty-four radiologists looked directly at the gorilla, yet 83 percent of them didn't see it. It's easy to imagine that if you're watching intently during your meeting to see who is and isn't taking notes, you could miss things like your boss asking you a direct question ("Bueller? Bueller?"). Would you notice the gorilla? If your attention is focused elsewhere, don't be so sure.

Well then, what happens when you are unsuccessful at keeping distractions from intruding on your attention? A fascinating line of research starting with a classic 1974 paper in the journal *Sociometry* shows that if you are distracted while listening to propaganda, you are more likely to believe the propaganda. It takes your full attention to evaluate a persuasive message, and if you're assigned a task along with listening, studies show that not only will you be more persuaded, but you will come up with fewer counterarguments and will remember less overall about the persuasive message; distracted, you accept the argument, sheep-like. Now think about *this* in terms of your meeting. If you're distracted, you're more likely to drink whatever Kool-Aid is being poured.

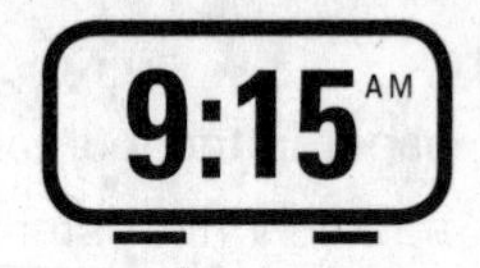


## *FOURTEEN WAYS YOU ARE A PUPPET TO YOUR UNCONSCIOUS*

Your brain is pushed and pulled by "priming" from the time you open your eyes until the time you close them. So you might as well learn these tips and tricks early in the day, so that you can watch out for them, rather than later in the day when you've already been led by your subconscious into the trap of illogical actions and decisions. It's time to take a look at how things below your consciousness affect what you think and do—how those little, seemingly inconsequential things make profound differences in your experience of the world. Here is a sampling of neat findings from priming studies. Do with them what you will.

• What you read in your morning media or listen to on the way to work influences how you act during the day. For example, after college students read sentences that included words like *gray*, *Florida*, and *bingo*, priming them to think of older people, they walked measurably more slowly down the hallway after leaving the lab.

• Big numbers make you expect big numbers. For example, if you hear a high estimate of a country's population, the price of a smartphone (thank you, Steve Jobs), or a person's age, your own estimate of population, cost, or age is likely to be higher.

• When you're frustrated at work, imagine a conversation with your significant other. Priming studies show that imagining the conversation will help you offer wiser solutions to problems.

• When the barista hands you a cup of coffee, what's the temperature? Studies show that people holding warm drinks evaluate strangers more positively than do people holding cold drinks.

• Do you see sick people? When a person sneezes, it primes you to worry about mortality—even unrelated health risks like heart attack and crime. And blood samples show that seeing images of sickness may actually prime your immune system to fight disease.

• If you need creativity, try standing next to a cardboard box, which can prime you to think "outside the box."

• Sometimes you need to act happy or mad. In that case, prime yourself by recalling memories of times in which you felt these ways. Keeping an "emotional priming Rolodex" of memories at hand can help you find the emotion you need in the moment.

• Beware the priming influence of stereotypes that you represent. For example, when Asian American women were primed to identify with the Asian stereotype, they did better on a mathematics test, and when they were primed to identify with stereotypes of women, they did worse on the test. Likewise, female students at a liberal arts school did worse on a test of spatial abilities when a short questionnaire primed them to remember gender stereotypes. It takes work and consciousness to avoid the magnet of your stereotypes.

• Need to create a simpatico connection? You can prime it with body language: he scratches his ear, you scratch your ear; she shrugs her shoulders, you shrug your shoulders. Mirroring a conversation partner's body language has been shown to increase rapport and connection. (That is, unless you get caught.)

• Big brother is watching. If your workplace shares something on the honor system—say, a snack bin or a mini-fridge—consider hanging a picture of eyes over the stash. A study found that when a picture of eyes was placed over a box where people paid on the honor system for snacks, money tripled; when a picture of flowers was placed above the box, people filched. Maybe foliage conceals?

• Your environment influences you subconsciously. In one example of many, when subjects read words that were flashed on computer screens and then read descriptions of a character with ambiguous behavior, people who had read hostile words like *hate* and *punch* were more likely to see ill intent than people who had read neutral words.

• Smile. How you use your body can influence your mind. For example, a famous experiment asked subjects to hold a pencil in their mouth while listening to a speech. With your mouth forced into the position of a smile, you will be more amused. If you need to feel smiley, fake it until you make it—smile and you'll feel better.

• Hack the break room. In one study, people who

smelled a cleaning product and then ate a snack of crumbly biscuits were more likely to clean up after themselves. Consider how you can prime the behavior you want.

### How to Manipulate Your Significant Other's Unconscious

Oh, gosh, the possibilities are endless and delectable! Can you prime your partner for better personal hygiene habits by putting soap on the aromatherapy candle? Can you prime the same person for a better evening by meeting him or her at the door with a warm drink? Taken one by one, priming experiments are neat; taken together, they show that we affect our environments and our environments affect us much more than you could ever imagine. In addition to using priming for your evil purposes, notice today how you are being intentionally and unintentionally primed for all manner of opinions and behaviors.

## *IF YOU BELIEVE YOU HAVE MORE WILLPOWER, YOU HAVE MORE WILLPOWER*

Can you resist the lure of the midmorning trip to the vending machine? What about that jar of candy you *know* is sitting on that intern's desk? Or in the middle of putting together that report, can you keep yourself from punching over to that open browser window to check celebrity gossip?

Willpower is your brain's ability to maintain attention even in the face of temptation, distraction, or discomfort. There are two schools of thought about willpower. One is that if you dig deep, you will always be able to find a bit more. The other is that when you run out of willpower, you are out, and at that point you need to do something to replenish the store.

Research supports both of these beliefs. Or, more precisely, some researchers believe that willpower is an unlimited resource and others think that it is a limited resource. One thing everybody agrees on is that willpower is a good thing. It's been linked with stuff including "healthier interpersonal relationships, greater popularity, better mental health, more effective coping skills, reduced aggression, superior academic performance, as well as less susceptibility to drug and alcohol abuse, criminality, and eating disorders," as Roy Baumeister and his Florida State

University colleagues write in the journal *Personality and Individual Differences*.

Baumeister happens to be in the "finite resource" camp. And while willpower seems like it might be one of those skills, like creativity, that floats around in the human brain blowing raspberries at researchers' attempts to pin it down, Baumeister believes that willpower is a whole lot more mechanistic than that. To him, willpower is all about sugar.

Do you remember back to this book's section about sugar and a healthy breakfast? The slow drip of glucose in your brain becomes especially important when you get to school or to work and are faced with the choice to actually get things done or to surreptitiously surf celebrity news on the *Huffington Post*. Most of the brain continues to function moderately okay as glucose gets a little low or a little high. Not willpower. Baumeister shows that when the going gets tough, the tough require sugar—scores on tests that require sustained attention (i.e., willpower) plummet when brain glucose is low, and scores on difficult (but not mundane) driving simulation tests are lower when glucose is lower. In fact, in a series of experiments Baumeister lays out a pretty persuasive case for glucose as the currency of willpower, summing it up like this: "First, measurements of blood glucose showed significant drops following acts of self-control, primarily among participants who worked hardest. Second, low glucose after an initial self-control task was linked to poor self-control on a subsequent task.

Third, experimental manipulations of glucose reduced or eliminated self-control decrements stemming from an initial self-control task."

In other words, self-control burns glucose; once available stores of glucose are burned, there is less self-control; and giving glucose to someone who has burned his or her store reinvigorates self-control. If you are tempted to check the celebrity news instead of working, drink a Tropicana Twister (4.38 grams of sugar per ounce!).

Then there's the other camp—the researchers who believe that willpower is an unlimited resource and that even without the influence of syrupy drinks you can dig deep to find more. Finding researchers publicly playing out a difference of opinion through competing journal articles is like science gold. And it's hard to read the following statement by a Stanford research group headed by Carol Dweck and published in 2010, just after Baumeister's article, as anything but a gauntlet lobbed in the general direction of the Florida group: "Much recent research suggests that willpower—the capacity to exert self-control—is a limited resource that is depleted after exertion. We propose that whether depletion takes place or not depends on a person's belief about whether willpower is a limited resource."

It's a nifty assertion, but where's the beef? The "beef" is, like Baumeister's claims, shown in a series of experiments. In the first, Dweck and her Stanford colleagues showed that for some people, willpower is depleted after a brain-draining task, whereas others can go on to the next challenge with no willpower fatigue. The difference between these two groups is what they believe. For people who believe that willpower is limited, willpower is limited, but for people who believe that willpower is unlimited, it is unlimited.

The second study manipulated what people believed about willpower and then tested them. Again, people taught to believe in the unlimited resource theory could always find more willpower, whereas people taught that the tank could run dry burned out after the first difficult task.

Dweck's third experiment showed that people in both camps found this task equally exhausting—it's just that people who considered willpower unlimited didn't let it exhaust them.

Finally, the fourth study looked out into the world at the lives of people with these two theories of willpower. A Web-based questionnaire assessed students' opinion of willpower and asked them to list a personal goal. Then later, during exam week, the questionnaire asked how often students had watched TV instead of studying, how often they had consumed high-fat or high-sugar foods, and how they had progressed toward their personal goal. You can probably guess the punch line: students who believed that willpower is finite had eaten more junk food, watched more TV, and worked less toward their goal than people who thought willpower is infinite.

Now that you've seen both arguments, which do you believe? Do you believe Dweck at Stanford or Baumeister at Florida State? Is willpower limited and based on glucose, or is willpower limited only by your belief? If you need to buckle down and finish that project while inhibiting the urge to check celebrity gossip, do you think it's better to dig deep or drink a soda? If you believe you can always find more willpower, you'll find it. If not, boosting your willpower may require sucking down a Pepsi and dealing with the consequences. Your waistline will thank you for trying the first strategy first.

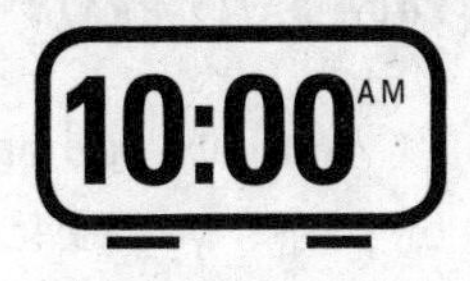


## *HOW TO STEP UP, NOT FADE AWAY, IN THE FACE OF WORKPLACE STRESS*

Here's a nice little finding about stress from our friend the juvenile rainbow trout. Apparently in "juvenile rainbow trout of similar size and with no apparent differences in social history," the ability to win fights for social dominance is greater in fish that shrug off the effects of stress, according to a study published in *Hormones and Behavior* in 2003. For your average juvenile rainbow trout with no especially distressing social issues, think of kicking ass as a job. And your average trout does its job better when it's not stressed.

You can probably see where this is going. But wait! If you are not a stressed trout but a stressed male ant, the effect of stress is flipped! In a lovely article titled "Mating with Stressed Males Increases the Fitness of Ant Queens," researchers show that mating with winged males—who tend to be totally stressed out compared with wingless ants—"positively affects life span and fecundity of young queens," which you have to admit is a bonus any way you look at it.

So here is the question: are you a trout or are you an ant? When the stress hits the fan at work, do you respond positively or negatively? When you are under pressure, do you fail to kick ass like a trout, or do you increase the life span and fecundity of the queen?

Actually, you probably do both, depending on the level of stress you're feeling. Most humans have an inverted U-shaped response to stress in which they perform crappily with no stress or with too much stress, but are at their best at the Goldilocks level of stress, where it's neither too much nor too little. That makes sense: stress puts some pep in your step, but too much can bury you. The thing is, you can manage your experience of stress to make a bigger Goldilocks zone. Here's how.

First, chronic stress is bad. You can only hum along at a high baseline level of stress for so long before it catches up with you. The Mayo Clinic notes that chronic stress puts you at higher risk for anxiety, depression, digestive problems, heart disease, sleep problems, weight gain, and memory and concentration impairment. There's even some evidence that stress may account for some of the health effects of living with a lower socioeconomic status. A paper from the University of Pittsburgh Cancer Institute points out that people with lower socioeconomic status tend to live in conditions that contribute to chronic stress, including "crowding, crime, noise pollution, discrimination and other hazards," and that this higher baseline stress that comes with being poor contributes to the health problems that tend to go along with poverty. No matter how you slice it, chronic stress is neither motivating nor productive, and eventually it leaves you feeling like a quivering little petri dish of *E. coli* bacteria. Really, if you're trying to be productive, healthy, or both, you gotta chill out the chronic stress.

But what about the thing that researchers call "acute"

stress—a project or unexpected challenge that takes you from a stress level of zero to sixty, but then allows you to go back down again? For example, researchers have studied this acute stress reaction by throwing rats in a pool of cold water. A classic study of the human response to stress comes from a surprisingly similar circumstance—Pennsylvania small-business owners thrown into the cold waters of a flood. Researchers interviewed 102 small-business owners after Hurricane Agnes swept through central Pennsylvania in 1972, and they found that on the Subjective Stress Scale, which runs from 0 to 100, a level of perceived stress between 40 and 48 predicted the highest performance—business owners in this range got the most done. But here's another interesting point: business owners' *perceived* level of stress had nothing to do with how much they lost in the storm. The level of stress they felt was completely independent of whether they had lost 5 percent or 100 percent in the flood. What this means is that stress, and the ability to keep stress in the productive Goldilocks zone, may not necessarily depend on what is done to you, but on how you evaluate and transform it. When it comes to predicting how well you will perform, what happens to you isn't as important as how you perceive it.

One secret of people who perform under pressure is the ability to manage the pressure. Can you reframe meaningless situations to be at least moderately stressful so that you can harvest the motivation that comes with a little stress? On the other hand, can you dial back your experience of mind-crushing stress until you perceive it as motivating but manageable? If so, you can learn to perform under pressure.

This study also shows that no matter how much stress you're under, how you cope with stress matters. Specifically, the study finds three general styles of coping with stress: people who deal with the stressful situation, people who deal with the emotional experience of stress, and people who use both coping strategies. First, the obvious: people who deal with both the situation and with their emotional response to stress perform the best under pressure. But check this out: people who cope with the emotional experience of stress do awfully until the going gets really tough, and people who deal with the situation are strong until stress gets high and the situation finally overwhelms them. As a stressful situation grows past the point where you can reasonably control the situation itself, it becomes more and more important to control your *emotional* experience of the stressful situation.

## Coping with Stress Done Two Ways

Next time you're supremely bored or ultimately stressed, try to keep the "inverted U" shape of stress and performance in mind: a little is good, but a lot is bad. Then if you can't help but exist at that point on the curve where stress threatens to overwhelm you, try to round out your natural coping tendency with strategies from the other side of stress management. If you're a doer, try managing your emotions; if you're a de-stresser, try resolving the situation that creates stress. By chipping away at stress from both the situational and emotional sides, you can give yourself the best chance at managing it effectively.

## *SHOULD YOU PAINT YOUR OFFICE RED OR BLUE?*

Check this out: a 1996 study of placebo pills published in the *British Medical Journal* shows that red placebos are more effective as stimulants and blue placebos are more effective as sleep aids. Another one: a study of Greco-Roman wrestling at the 2004 Olympics shows that wrestlers randomly assigned red rather than blue uniforms won 55 percent of the matches. More? There's evidence that blue outdoor lighting may reduce crime.

You've heard it before, and science bears it out: color influences your mood. Heck, color influences you even when you aren't aware that you've *seen* the color. For example, in a study reported in *Psychological Science,* when green, blue, or white discs were flashed, even when subjects weren't aware of the color they had seen, it affected their later perceptions of other colors. In this experiment at least, it's not as if the unconscious perception of color turned pussycats into tigers, but it shows that you don't have to be aware of color for it to affect you.

If you believe the traditional view that our ability to see color evolved so that we could figure out which fruit is ripe, the effect of color on our emotions doesn't make much sense. Why should an unripe fruit make us calm and a ripe fruit make us agitated? But what if the human ability to sense colors with our eyes evolved not to gauge the color of fruit but to help us read emotions? See, it turns out that the color-sensing cones in our eyes are exactly optimized to detect changes in color due to blood oxygenation—the red of oxygenated blood and the blue of blood carrying less oxygen through the bloodstream. And the primates who have evolved the ability to see color are also the primates whose hairless faces and/or rumps let us display emotions through our blood-rich skins.

While it's important to know which fruit is best to eat, it may be even more important to know the emotions of those around you while you're eating. And once millions of years of evolution trained the human brain to see red as aggressive and blue as calming, it's not as though a puny little thing like your cognitive mind can just override it. Now when a wrestler sees his or her opponent in a red one-piece, it seems as if that opponent is aggressive and angry. Maybe wearing a red one-piece even *makes* that opponent slightly more aggressive and angry. Because the basic biology of your eyes is optimized to recognize these colors and your brain has evolved to interpret them, you may feel just a touch meeker standing there in your blue suit. Or you might expect that a blue pill will make you feel as if your blood is deoxygenated and a red pill will make you feel full of pepped-up hemoglobin.

Or you might find yourself stressed and steaming in a red office, whereas you could find yourself calm (and sleepy?) in a blue office. The question of what color you should paint your office comes down to how you need to feel while at work. Paint it red if you need aggression. Paint it blue if you need to chillax.

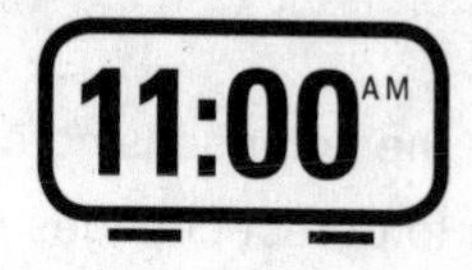


## IF NOT IQ, THEN WHAT? PRACTICAL INTELLIGENCE HELPS YOU SURF WORK DILEMMAS

One huge reason experienced employees are so valuable to an organization is that no matter how smart you are—or how smart that whiz kid straight from the Ivy League with a computer science degree appears to be—the job-specific practical intelligence that comes from experience isn't something you can teach. As an employee, you know things that you don't know you know—all the assumed, background knowledge that allows you to smoothly navigate situations, even new situations, without being told what to do. Researchers call this *practical intelligence*. On the opposite end of the spectrum from your Oliver Twist–like practical intelligence is a computer (and that annoyingly peppy whiz kid). For example, imagine you're telling a computer how to make a peanut butter and jelly sandwich. In your human brain, you might understand that when you say "spread jelly," you intend it to be done with the help of a knife or at the very least a spoon, but the computer does not. It has none of this knowledge that we assume to be mutually understood. Once the jelly is spread, you have to tell the computer to put the jelly slice and the peanut butter slice together so that the jelly comes into contact with the peanut butter; otherwise you might end up with an inside-out sandwich.

How many people do you know who are like this

computer—with tons of computing power but without the practical intelligence to use it to their advantage? C'mon, we *all* know somebody. Or maybe this is one of those look-to-the-left, look-to-the-right moments and it's *you* who has difficulty navigating the niceties of situations that others seem to skate through with ease. We all have a different amount of overall practical intelligence. As with IQ or creativity or height, you hold within you a measurable quantity of practical intelligence that makes you know how to handle things or not know how to handle things.

Each person also carries practical intelligence that is specific to situations. You may know that in the culture of your specific cluster of cubicles it's okay to talk in ribald ways across the dividers about what did and didn't happen this past weekend. Or you might know that in order to trim the number of unnecessary department meetings, it's most productive to chat in the break room with the boss's secretary, who will filter the opinion to the higher-ups, instead of going to the boss directly.

This is practical intelligence. And whereas intelligence depends on your conscious mind, practical intelligence depends in large part on your unconscious mind. Instead of intentionally learning things, you happen to pick things up; instead of learning explicitly, loading your practical intelligence depends on learning implicitly.

To speed it up, try moving practical intelligence from the unconscious to the conscious mind. Throughout this book, we'll try to stay away from catchphrase movements that haven't been through (or haven't stood up to) the pokings and proddings of science. That said, it's sure looking like *mindfulness* is more than a California/yoga/sushi fad. Basically, mindfulness is the difference between just sitting in front of the TV eating a package of Double Stuf Oreo cookies and sitting in front of the TV eating cookies while being aware of what you are doing and why. It's not that mindfulness forbids or ensures any specific behavior; it just means that you can be less led by the ring in the nose of your subconscious into doing, thinking, feeling, and believing things that your conscious self would rather not.

When you examine your actions instead of simply performing them, you have the opportunity to bring to the foreground of your mind things that would usually be left in the background. You have the opportunity to make implicit things explicit and to use your intelligence to boost your practical intelligence. You want to learn how? Keep reading.

## Mindfulness and Practical Intelligence

Now we're in the weeds of interconnected things that are given different names but are all mushed up together, including practical intelligence, emotional intelligence, cultural intelligence, and more. For every one of these somewhat imprecise terms, there's a study showing that mindfulness trains it. Here's one way to practice. Ask yourself what background knowledge a computer would need in order to accomplish one of your tasks at work. What would the computer need to know in order to talk with your nonprofit's donors? Or your boss? Or the little people? What is the work equivalent of knowing to put the peanut butter side of the bread in contact with the jelly side of the bread? And then, just as important, ask *why* that's the case. By using mindfulness to make your implicit learning explicit, you can harvest more practical intelligence from these situations that require knowing without knowing.

## *WHAT IS YOUR LEARNING STYLE?*

Gosh, you've heard so much about learning styles—some people learn best by reading, others by listening, some by doing, yet others by watching, and so on . . . or so the story goes. For some of us the following is tough to stomach: when researchers have looked behind the curtain of learning styles, instead of the magic of the Great and Powerful Oz they've pretty much found a short traveling snake oil salesman from Kansas pulling levers and making grand pronouncements. Sure, people know which learning styles they prefer. But there's no evidence that actually matching the style of instruction to a person's preferred learning style makes him or her learn any better. For example, a 2012 report commissioned by the U.S. Department of Education observes, "In short, there exist a smattering of positive findings with unknown effect sizes that are eclipsed by a much greater number of published failures to find evidence, and we suspect that additional null findings sit in researchers' file drawers."

What this means is that when you take a single subject, say math, and teach it in a visual way to visual students, an auditory way to auditory students, and a kinesthetic way to kinesthetic students, it hasn't been shown to create better results than using one method of instruction for all students. In fact, when you try to teach math through kinesthetics or art by talking about it so that instruction can match students' preferences, learning suffers. That's because, as much as some parents may wish otherwise, movement is not the best way to teach math, and art is a visual exercise, not an auditory one.

Instead of working to match instruction to students' learning styles, the Department of Education report suggests that "educators should focus on developing the most effective and coherent ways to present particular bodies of content, which often involve combining different forms of instruction, such as diagrams and words, in mutually reinforcing ways."

The same is true for you at work. What matters most is not necessarily how you like to learn new skills and information, but how the particular skills and information are best presented. As much as you fear Excel, Photoshop, or the interpersonal mechanics of learning a new skill directly from someone who already knows it, if it's the best way to get the information, it's the best way to get the information, regardless of how you think about your learning style.

## The Research Cost of Learning Styles

Daniel Willingham, professor of psychology at the University of Virginia and leading debunker of the learning styles myth, points out that not only might belief in learning styles doom a "kinesthetic learner" to inefficiently trying to learn math by skipping rope, but the research invested in learning styles takes away from other, more important, and more promising things that people could be studying. Couldn't all those education researchers' crackling neurons be put to better use solving actual problems in learning and school systems? Your assignment today is to stop believing in learning styles. Just stop.

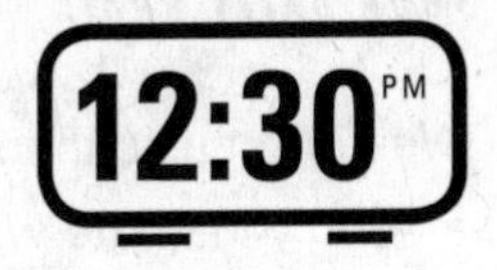


## *HOW TO MAKE A CREATIVE IDEA POP INTO EXISTENCE*

According to Gordon Kane, director of the Michigan Center for Theoretical Physics at the University of Michigan at Ann Arbor, teeny-tiny quantum particles do indeed pop into and out of existence. They spring into being from nothing, "temporarily violating the conservation of energy," Kane writes in an article for *Scientific American*. It seems like that's what a creative idea does too: it springs into existence from nothing. But that's not the case at all. Instead of a lightbulb popping into being above your head, creativity builds this lightbulb from spare parts it finds lying around your brain: the glass from here, the filament from there, the threaded metal cap from another place.

Instead of generating something new, a huge field of research shows that creativity comes from making new connections among information and ideas that already exist in your brain. In other words, it depends on something called *domain-specific knowledge*. In order to be a creative electrical engineer, you need to know stuff about electrical engineering; in order to be a creative artist, you need to know things about art. Think of each "thing" you know as a little dot on a grid. The more dots, the more ways you can connect them. As YouTube videos show, a cat walking on the piano keyboard can break the rules of

Western music theory, but it takes knowing these rules in a deeply intimate way to break them in a way that is truly *creative*.

Now, the real challenge of creativity is gathering these dots of information without becoming bound by how they are usually connected. For example, as you learn about painting, you will learn that a painter uses a brush in a certain way to spread paint on a canvas. Bound by this connection, you would never become Jackson Pollock. (Interestingly, once Pollock's creativity severed this connection between painting and using a brush, there was an explosion of painters printing, spitting, spraying, or otherwise introducing paint to canvas without the traditional application of a brush.)

One way to gather knowledge without being bound by it is to gather it from many fields. Take the famous example of Pete Gogolak, which you undoubtedly already know. Gogolak grew up in Hungary playing soccer. Then he ended up playing American football as a student at Cornell University. He could kick, and so he became the kicker. Only he didn't kick like a football player; he brought knowledge from another field—the soccer field—and instead of approaching the football from straight behind, Gogolak started approaching the ball from two steps back and three steps to the side so that he could kick it like a soccer ball. Gogolak went on to play for the New York Giants in the 1960s and 1970s, and by the end of his career almost everybody was kicking field goals soccer style. He had the domain-specific knowledge of kicking but connected it to American football in a new way.

This infusion of creativity from other fields is why students at MIT might do well to take classes in the humanities. To be a creative electrical engineer, you need the domain-specific knowledge of electrical engineering, but then it helps to have other knowledge from outside the field to infuse into it. Limiting the breadth of your knowledge limits the depth of your creativity.

You also need to be able to hear your creativity. The thing about making new connections in your brain is that they're rarely as loud as the shouts from the established connections. The way things have been done in the past tends to drown out the way things *could* be done in the future. You have to listen closely to hear creativity over the din of your experience. Experiments show that the more experience, the louder the din. We've already seen one way to do this: hit your snooze alarm. A brain on the cusp of sleep or in an unfocused, relaxed state is largely cleared of preconceived thoughts and ready for the burst of creative insight. Likewise, a classic article in the *Journal of Creative Behavior* followed Cornell undergraduates (no mention of Gogolak, though) for years after they learned the technique of transcendental meditation and showed that students who kept up the technique were able to use it to generate creative thoughts.

Let's be honest: "turning the attention inwards toward the subtler levels of a thought until the mind transcends the experience of the subtlest state of thought and arrives at the source of thought," as recommended by Maharishi Mahesh Yogi, probably sounds a little weird. And if it sounds a little too weird to try, chances are you'll never be quite as creative as you could be. That's because a final ingredient of creativity is the piece of your personality that determines how bound you are by cultural norms. Would you be terribly mortified if your coworkers or classmates saw you sitting barefoot and cross-legged, deep in meditation? Would you ever model nude for a life drawing class? Could you imagine thumbing your nose at the tradition of art that stretches back to fourteenth-century Florence and flicking paint at a canvas?

If you have been diligent in your acquisition of domain-specific knowledge, have looked beyond your field for influences, are willing to listen to whisperings from the back of your mind, and are predisposed (or can find the courage) to shake things up, you're primed to take the next great leap in creativity, be it in electrical engineering, art, or football kicking.

## Is Creativity Thinking *Inside* the Box?

A 2004 review in the *Creativity Research Journal* looked at the lessons of seventy studies to show that creativity can, in fact, be trained. There are things you can do that will make you more creative. The review also shows what works: namely, cognitive training is better than training that tries to hit creativity from social, personality, or motivational angles. This means that boosting creativity requires changing the way you think, not necessarily adjusting your mood or motivation. To make your thoughts more creative, focus on identifying problems, generating ideas, and combining concepts in new ways—the review shows that training in these areas offers the biggest boost. Interestingly, it wasn't the "divergent thinking" skill of brainstorming that led to the biggest creativity boosts. Instead, it was skills involved in analyzing problems, such as convergent thinking (defining answers), critical thinking, and constraint identification. If you want to create creative solutions, first define what you know and what you can (and can't) do, then practice working within these parameters to make new things from old information.

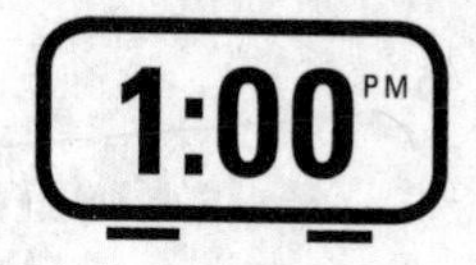


## *SHOULD YOU DO TODAY WHAT COULD BE PUT OFF UNTIL TOMORROW?*

Today is the day you finally buckle down and organize all the files in your department's shared drive. Or go to the campus library to check out books on your research project about ancient Mesopotamia. Or clean the leaves from your gutters. I mean, you know that disorganized files cost you a couple of minutes every day, that eventually you will have to learn about ancient Mesopotamia, and that if you leave the leaves through the next major rainstorm your gutters are likely to rip from the roof, potentially crushing Girl Scouts who have come to your front door to sell cookies, thus depriving you of cookies.

But you would spend more time today organizing the files than you would spend searching through them, your research paper isn't due until after the weekend, and there is no rain in the forecast this week. And so you give in to what an article in the journal *Psychological Bulletin* calls "a prevalent and pernicious form of self-regulatory failure."

Damn, that seems kind of harsh. How bad is procrastination, really?

First, it's bad enough that when you try to study it, you find things like this: "High scorers on the procrastination scale were more likely to return their completed inventory late," from an article in the *Journal of Research in Personality*. Who are these slackers whose very nature makes them difficult to evaluate in a consistent way?

They are largely the ones you would expect: disorganized, impulsive, distractible people who are likely to rate their enjoyment of projects higher when the time it takes to complete projects is lower. At least that's the case when the project is assigned by some force beyond themselves. See, procrastinators also believe in their own self-efficacy and are motivated by factors other than achievement.

More and more, research is showing that procrastination isn't a defect in ability but rather a disconnect between the demands of a task and what motivates the procrastinator. Procrastinators are intrinsically and not extrinsically motivated, meaning that neither tempting them with rewards nor warning them the sky will fall is likely to up their motivation to the threshold of action. Instead, the procrastinator has to *want* to do something. Maybe he or she would start this minute on a model of an ancient Mesopotamian ziggurat, but no amount of threatening or cajoling will make that person write a report on the same thing.

If *you* are the procrastinator (forsooth!), the secret to reducing your procrastination—something that 95 percent of procrastinators claim to want—is not to focus on rewards and punishments or to create new rewards and punishments, but to find aspects of a project that motivate you. No one else's reasons will make you get to work, but discovering the aspects of a project that ring true to you will help you get started.

According to the authors of the *Psychological Bulletin* article, one thing is certain: "Further research on procrastination should not be delayed."

## The Carrot and the Stick

If intrinsic motivation for a project is just too hard to find, consider another mind trick to reduce procrastination: reward yourself for finishing. This is the "get your work done so that you can play" strategy. Placing a motivating task at the end of a non-motivating task can make you work to remove the block. That's the carrot approach, but there's also the stick: you can try imposing costly deadlines on yourself. For example, if you can't finish by a set time, you will give $5 to a political cause you disagree with. But an article by Dan Ariely in *Psychological Science* shows we're pretty bad at setting these self-imposed deadlines. When he paid students for proofreading, students who self-imposed deadlines (as opposed to working with an externally set deadline) found fewer errors and earned less of the researcher's money. Again, people who are predisposed to procrastinate do best when they can present themselves with an internal carrot, rather than when they are presented (or present themselves) with an external stick.

## *PACK YOUR WORKING MEMORY*

If you have ever packed a backpack for an overnight hiking trip, you know that everything has its place: your sleeping bag sits inside the zippered flap at the pack's base; the heavy food bag sits low and tight against your back; your foam sleeping pad is strapped to the outside, and the inflatable pad that sits on top of it is shoved inside; little stuff sacks hold your cooking gear, extra clothes, and toiletries. All that stuff is packed away for when you need it later in the trip. But your backpack's top flap, ah, that's where the important stuff goes, everything that you need during the day—your snacks, sunscreen, map and compass, roll of duct tape for blister care, and UV water purification pen. Tomorrow you may move new food from the packed-away bag into your top flap, but for now you have everything you're likely to need accessible right there on top.

Your brain is the same way. You have many kinds of memory, each with its place. Kinesthetic memory is different from emotional memory, which is different from your memory of faces, which is different from the storage of facts. The vast majority of these things we leave packed away deep in our brains. The stuff we really need right now is what we move to our brain's top flap: working memory. If you want to be able to work with more information, you'll have to expand the capacity of this top flap that is your working memory.

Because working memory is where we manipulate information for any purpose you can imagine, the power of your working memory is absolutely central to almost everything you do with your brain. We've seen that if working memory is plugged with stereotype threat, you may crash and burn when parallel parking or taking a math test. And the limit of your working memory is part of why you can't multitask—you just can't hold that much stuff in the front of your mind.

That's because it really doesn't hold much. The mistaken opinion is that your working memory can hold seven "chunks" of information, and you may have heard the mistaken extension of this opinion that your seven-digit working memory is why we have seven-digit phone numbers. We know now that very few of us have a full seven chunks but we also know that with practice, knowledge, and expertise, we can come to hold a vast amount of information inside a single chunk. Here's an easy example: if you are trying to remember the phone number of someone who lives in southern Ontario, holding on to the area code 519 may claim three full chunks of your working memory; but if you're trying to call a friend in the town of Reeves, Louisiana, it would certainly only take one chunk to remember 666. (Or it used to: the residents of this small, religious town successfully petitioned the federal government to decouple their town from the number of the devil, but that's beside the point.) Likewise, you would have a hard time remembering the letters *m-a-s-c-a-r-p-o-n-e*, but putting them together into the word *mascarpone* (especially if you have expertise in Italian spelling) would claim only one chunk or less of your working memory.

Chances are you're already an expert at remembering a number with repeating digits or letters that form words. If you were a chess expert, a pyramid of pawns might be kind of the same: you would look at it, immediately recognize and categorize it, and assign the entire pattern to one chunk in your working memory. If you think that chess is played by taking your opponent's pieces, you might have to remember each of these pawn positions individually, using much more working memory. Something similar is true of any kind of expertise—the more expert you are, the more you are able to "big-chunk" an assortment of bits of information that you know are related.

Think about it. What do you know? That is, in what areas of your work, hobbies, family, or arcane knowledge are you an expert? If you know baseball, you'll be able to "big-chunk" the entire sequence of an inning; if you know how to bake a soufflé, the list of ingredients may not even claim a full space; if you know how to build trestles, you might not need to painstakingly write down every measurement. But beware your expert attempts to categorize and chunk. Another feature of expert working memory is the tendency to assume that every new case is the same as the things you already know, which means that experts are prone to overlooking things. You can see this in chess. There's a ranking at which chess players are very good at mining their expert working memories for moves to match most board positions . . . and then there's an even higher ranking in which masters know the "right" moves but aren't bound by them. That higher ranking is the pinnacle of working memory.

## Biggering Your Brain

Researchers have shown that repeated practice with brutally difficult brain games can grow your working memory, and one side of a raging debate in the field of intelligence research even believes that training working memory can boost your IQ. One promising brain-training game, called the "*dual n-back*" asks you to remember strings of information—not just a string of numbers written on a sheet of paper or a string of letters read through headphones, but both at the same time. You monitor these strings, and then you report the letter, number, or symbol that you heard two-ago or three-ago or four-ago (thus the *n* in "*n*-back"). For you, the *n*-back is only a quick Google search away. If you really want to do something powerful to bigger your brain and you aren't afraid of hard work, this training can expand the area of your brain where you get things done.

## IT'S ONLY TWO O'CLOCK. ARE YOU BURNED OUT ALREADY?

Here you are in the final third of a long workday, at the end of a long week, at the end of a long year . . . at the end of your rope. You've been pouring yourself into making widgets, and for what purpose? You're exhausted, more emotionally than physically, but your lack of emotional energy makes every action feel as if you're pushing through a thick vat of molasses. You couldn't care less about these damn widgets. In fact, you feel like if you have to see one more today, you're going to chuck it off a dock.

This is understandable, right? We all get tired of the things we work with, the things we make. But what happens when a "widget" is a person? Burnout can be equally if not more fierce in people professions. And in this case, the consequences may go beyond making a bad widget and into the realm of emotional or physical danger for the people who live on the other end of your burnout.

The classic example is nurses. First, there's the shortage thing: each additional patient added to a nurse's workload has an additional 7 percent chance of dying in the thirty days after admission. And each patient added to a nurse's workload also increases the chance of professional burnout by 23 percent. Sure, it's harder to juggle seven balls than it is to juggle six. But in addition to the difficulty of watching over more people, a major component of nurse burnout is cynicism and depersonalization—not only do nurses (and other service professionals) start to see people as widgets, but they start to feel cynical about patients' motives, pessimistic about their skills, and start to believe in the inevitability of dire outcomes.

"This callous and even dehumanized perception of others can lead staff to view their clients as somehow deserving of their troubles," says an article in the *Journal of Occupational Behavior*.

There have been many solutions posed to the problem of burnout, ranging from more time off in order to recover emotional resources to shifting from one position to another within an organization to avoid the burnout of boredom and complacency. But if you really want to combat or cure burnout, the best thing you can do is find a way to increase your sense of personal accomplishment.

Really: The Maslach Burnout Inventory, the most common test of burnout and impending burnout, shows that on the other side of the depersonalization and emotional exhaustion that predict burnout sits personal accomplishment, which protects against it. It turns out that burnout necessarily includes the lack of personal accomplishment, and that "feelings of competence and successful achievement in one's work with people" can protect against burnout. If you can answer yes to questions like "I feel I'm positively influencing other people's lives through my work," "I have accomplished many worthwhile things in this job," and "I deal very effectively with the problems of my clients," then you are pretty likely to survive even an emotionally exhausting job. It's about making a difference. And if you haven't felt it in your work lately, it's time to find it outside your work, evaluate how you can do your job differently, or look for other gainful employment. On the flip side, sitting in the muck of ineffectiveness is a recipe for burnout.

**With a Little Help . . .**

Studies of nurses also show that focusing on the value of your peers can help you avoid burnout. It's as if recognizing that the people you work with are still people can help you hold on to the idea of your clients as people as well. "Through solidarity, caregivers will be better equipped to overcome the stress of caring for people," writes an article in the *Nursing Clinics of North America.* If you're burning out, get together with people in the same boat.

## *HOW TO READ YOUR COWORKERS' MINDS*

There he is, the hot coworker of your intimate dreams, eating a meatball sub while gazing down the hall toward your cube with a look of pure rapture. There she is, your boss, pushing her horn-rimmed glasses back up the bridge of her nose while eyeing the report you just landed on her desk. What do these people mean by these gestures? The secret is putting yourself in their shoes and trying out the same gestures to see what you would mean if you were doing the same things. Actually, your brain does this for you automatically.

We know because in the 1980s a group of Italian researchers inserted electrodes into the brains of monkeys to record the firings of neurons in response to the monkeys' motions. A monkey would move its arm, neurons would either fire or not fire, and through this painstaking process of probing and checking, the researchers went about mapping the homes of these movements in the monkey brain. Now, watching monkey brains is hungry work. And rather than taking time away from the task of monkey monitoring, researchers brought their food to the lab. You can imagine researchers staring at monkeys as the monkeys brought food repetitively from a dish to their mouths while the monkeys stared back at the researchers doing the same. Of course, the researchers saw this eating action in the monkeys' brains—hand to food, food to mouth, neuron fires, repeat ad infinitum.

Only it wasn't just when monkeys fed themselves that the Italian researchers saw these "feeding" neurons firing in the monkey brains. When the *researchers* ate, monkey brains were activated as if the monkeys themselves were being fed. You know the feeling of standing around in a group in which one person has brought a latte and the rest of you watch and drool? Your envy is driven in part by the fact that your brain is experimenting with what it would feel like for you to have a latte too.

When monkeys watch researchers eat, their brains mimic the action. When you experience another person, your brain can't help but mirror their actions and emotions, trying them out as if they were shadow actions of your own.

This mirroring is due to special cells called *mirror neurons*, which live in your brain but keep a close eye on the people around you. When someone smiles at you, your mirror neurons try out the smile in your own brain; when someone frowns or eats M&Ms, it's as if you frowned or ate M&Ms. Despite what you'll read elsewhere, that's about the extent of what we know for sure about mirror neurons: they exist, they mimic others' actions, and they're very, very cool.

There are three theories of why we have mirror neurons. Maybe we have them because in our evolutionary past, it was nice to know if Thog was suggesting that you work together to take down a mammoth or suggesting that he was going to end you in a messy way. Another theory is that we evolved mirror neurons so that we could learn from others' experience. You know that you learn from your own experience: the more you throw darts or paint brushstrokes or hit a tennis ball, the better you get at these things. But you can also learn by watching other people do these things, and perhaps mirror neurons allow you to encode others' actions more precisely in your brain. Maybe by automatically imitating others' actions, you get a trickle-down version of this training yourself. Finally, mirror neurons may allow you to empathize. When you say to a friend "I feel your pain," mirror neurons may make it literally true—and this mirror representation of another person's emotion may be why, when you're with someone who is anxious or depressed or elated, you can feel the emotional residue of their experience.

With this take on the role of mirror neurons in empathy comes a proposed link between the malfunction of mirror neurons and autism. This link even has its own catchy name: the "broken mirror" theory of autism. Still, for now there are compelling papers on both sides.

## Ape Me, Baby

In a fascinating condition called *echopraxia*, people compulsively imitate others' movements—what you see is what you do—and increasing evidence shows that the condition may be due to the inability to inhibit the urges of mirror neurons. When watching another person act, most brains reenact a version of that movement, but the brain is where it ends. In echopraxia, mirror neurons spark real movements. But we all have a shadow of this compulsion—when a conversation partner props an elbow on the table, you want to do the same (and the more empathically connected you feel with this person, the stronger your compulsion to copy). Next time you have a conversation, try to be aware of how your conversation partner's actions influence your own, and if you like, see if you can get the other person to copy your gestures or body language. The degree of copying can show you how strongly your conversation partner's mirror neurons are firing in your honor.

## *THE SECRET TO SOLVING ALL YOUR PROBLEMS*

Unless you are very lucky or very unlucky, your job and your life require very little problem solving. Think about it: most of the challenges you see regularly are problems you've solved before, and unless you're lucky enough to be in a profession that demands you generate new knowledge, or unlucky enough to be forced by circumstance to solve new problems that affect your life, you can pretty much coast on solutions you've generated in the past.

For the few new challenges you haven't faced, you can usually hire expertise rather than stumbling through the problem-solving process yourself. For example, imagine you're installing a new dishwasher. Unless you already know how to plumb from the garbage disposal, it may seem easier to work a little harder at your job to pay for a plumber rather than try to install the new dishwasher yourself. I mean, you could do it . . . but it would take some figuring out.

The rest of this entry is for those times you can't avoid figuring it out—and especially for when a solution doesn't come from rote and needs a little hit of creativity. Like the "innovation" angle of creativity we looked at before, creative problem solving remains a trick of connecting information in new ways, but there are some new strategies specific to this flavor of creativity that can help you pull it off. Here are some things to try.

First, try to avoid the temptation of the solution. It's hard!

I mean, that's what you *want*; why not just reach out and grab it? One major lesson from studies of problem solving is that the sooner you reach for the apple at the expense of understanding the ecosystem of the tree, the less likely you are to find a fruitful solution to your problem. In creative problem solving, it's all about understanding the problem. Specifically, it's about understanding what researchers call the *initial state* and *constraints*—in other words, being crystal clear about the problem, the rules, your resources, and the goals.

For example, take this classic problem used in problem-solving studies from the book *Lateral Thinking Puzzlers* by Paul Sloane: "Bombs Away: One night during the Second World War, an allied bomber was on a mission over Germany. The plane was in perfect condition and everything on it worked properly. When it had reached its target, the pilot ordered the bomb doors to be opened. They opened. He then ordered the bombs to be released. They were released. But the bombs did not fall from the plane. Why should this be so?"

Now work to understand the initial condition and constraints. Picture the problem in your mind and then ask yourself, from the information given, if your picture must be as you imagine it. Must the plane be flying? Well, if it's "over Germany," then it really must. Must the bombs be free to fall? Well, if they "were released," you would really have to assume so. Must the rules of gravity apply? Well, unless otherwise stated, you would have to assume the rules of physics are unchanged. Imagine the problem as if you were opening a hand that holds a tennis ball. Does the ball drop? Well, it does if your open hand is palm down. Do you see the solution now?

If you're like most people, you assumed something about the problem's initial state, but nothing in the problem states that the plane has to be flying right-side up. Of course, the solution is that the plane is flying upside down, and so when the doors are opened and the bombs are released, they don't go anywhere.

Let's strip down creative problem solving to a simpler form and try it again. Analogies have been extensively studied by people who study these things, and you've probably seen them before: how is A like B? Well, it's pretty easy to see that apples are to fruit the same way wheat is to grain (both are *kinds* of the category). But now try this one:

**MACERATION : LIQUID**

A. Trail : Path
B. Evaporation : Humidity
C. Sublimation : Gas
D. Erosion : Weather
E. Decision : Distraction

It's tempting to kind of unfocus your eyes and start putting answer choices next to the target and seeing how they *feel*. If you're totally in touch with your insight (see earlier entry on creativity), this might actually work. But if the insightful solution doesn't wallop you in a visceral way, you'll have to do some analysis. And by now you should know that in creative problem solving, that starts with understanding the problem: what are its initial state and constraints? The more time you spend working to understand how the problem works, the faster you will find a solution. So ask yourself: what is the relationship between *maceration* and *liquid*? Well, hopefully you know that *maceration* means "chewing," which makes you able to come up with the sentence "Maceration turns something into a liquid." (And if you don't, well, that's where knowledge can aid problem solving.) Now you can go to the answer choices. Does a trail turn something into a path? Kind of, but not really. Does evaporation turn something into humidity? Again, kind of, but not really. Does sublimation turn something into a gas? Again, solving this problem takes some knowledge—namely, knowing the definition of *sublimation*—but the answer is that heck yeah, it does. (Or you could have applied your practical intelligence to understand that the answer is always C.)

Let's try one more, again from Paul Sloane: "A man walks into a bar and asks the barman for a glass of water. The barman pulls out a gun and points it at the man. The man says 'Thank you' and walks out."

Instead of reaching immediately for a solution, work to understand the problem. Search for false assumptions. Might there be another definition of *gun*? (No.) Might "pointing" be something other than the obvious? (No.) What do you assume is the intention of the barman's actions? Do you assume that he is angry that the man ordered water? By prodding your assumptions, you may find possibilities in addition to the obvious ones. Is it possible the barman was trying to help the customer? If so, what problem would the customer think could be solved with a glass of water that could also be solved by looking down the barrel of a gun? Discovering the false assumption of the barman's anger leads you to the answer: the man had the hiccups, which the barman cured with a good scare.

## I-Ching Problem Solving

Problem solving is a big skill, and there are dozens if not hundreds of tips and tricks proven to increase problem-solving skills. One you may not have heard before comes from a 2013 article in the journal *Thinking Skills and Creativity*: applying the problem-solving strategies of the ancient Chinese philosophical text the I-Ching. The study looked at how 188 students applied I-Ching principles during the 2011 GreenMech challenge in which groups of four had to "assemble parts into a reaction system based on scientific principles and green concepts" (kind of like a crunchy Rube Goldberg challenge). In the BaGua model of the I-Ching, there are four paired problem-solving forces: sensitivity to the problem is balanced by arriving at a solution, mood arousal is balanced by logical reasoning, internal stillness is balanced by external action, and confrontation is balanced by support. The study integrates the BaGua with a traditional model of creative problem solving to suggest the following five stages of creative problem solving. With practice, you too can bring the ancient wisdom of the BaGua to your attempts at creative problem solving.

1. The process of sensing that a problem has occurred or will occur.
2. The activation of working memory during creative problem solving suggests that the idea-generating or -gathering processes are stimulated by mood arousal and logical reasoning.
3. The idea-generating and -gathering processes encompass individual interactions with people, data, and things, as well as meditative thinking during calm intervals.
4. The idea transformation process embraces two extremes to refine ideas: decreasing conflict about a new idea and deliberating about ideas.
5. Consequence-based idea evaluation is used to examine the effectiveness of different options.

## *HOW TO FIND WHAT YOU NEED ON THE INTERNET*

Name a small, noisy bird. *A chipmunk.* At what month of pregnancy does a woman start to look pregnant? *September.* What is a man's name that starts with the letter *K*? *Kentucky Fried Chicken.* Something that a burglar would not want to see when breaking into a house? *A naked grandma.*

These are real answers, folks, given by subjects in a study of free association published by the ABC Television Network under the title *Family Feud.* As you know, to play the *Feud,* a genial host asks contestants a question with an ever-so-slightly lewd undertone, at which point the pressured contestant tries to come up with the G-rated answer most given by one hundred non-pressured people who were polled ahead of time. It's the same skill that at work can make the difference between a five-second Google search and ninety minutes spent browsing for the information you need. Think about it. How many times in an average workday do you search for information online? And how did you learn to perform these searches? If you're like most of us, the answer to that last one is trial and error. Now's the time to bring just a little bit of science to your Internet search strategies.

First, there are a ton of search strategies to choose from, many of which you're probably already using without knowing you're using them. Take a minute to see which of the following strategies, described in a 2012 paper in the journal *Information Processing and Management*, you already use in your searches:

- Identifying search leads to get started (e.g., deciding to search CNN.com rather than all of Google)
- Creating a search statement
- Modifying the search statement (e.g., if "Why does my operating system hate me?" was too broad, you might try "Why does Mac OS 10.7.5 hate me?")
- Evaluating search results
- Evaluating individual items
- Keeping a record
- Going forward
- Going back
- Learning (e.g., learning how to use a more advanced search interface)
- Exploring
- Organizing (e.g., sorting results by date instead of name)
- Monitoring
- And, finally, *using*

Here's the important part: some of these strategies work better than others. We know what works because researchers have compared the behaviors of expert searchers to those of novices. Here's the scoop: expert Internet searchers spend much more time than novices evaluating search results, going back, and organizing search results—it's as if they do more with fewer search terms, getting intelligently into a "search ballpark" and then working systematically from there (as opposed to throwing many hopeful searches against the wall and seeing what sticks). Here's another cool little expert nuance: "evaluating search results" isn't clicking individual links to see if they give the information you want, but looking at the quality of the list of results as a whole. Did the search terms get you in the ballpark? If so, might you be able to tweak the search terms to narrow that ballpark? Finally, experts know not to chase white rabbits down holes—if a click takes them further away from rather than closer to their information goals, experts go right back to the most recent page that was on the right track. Don't be afraid to click that back button, and then stay organized in a way that keeps you from going down that dead end again.

Taken as a whole, the process of expert Internet search is much like the process of problem solving itself: an efficient Internet search requires time spent on the *front end* designing the problem-solving strategy, and much less time spent actually executing this strategy. Use your powers of *Family Feud* free association to search the terms that most people associate with your information. Tweak your terms to refine the results. And when in doubt, go back rather than forward. How many minutes or hours do you think you can save in the course of a forty-hour workweek by dialing in your powers of Internet search?

## Free Association Free-for-All

Do you want to test your ability to win *Family Feud* . . . and in the process get a little better at picking Internet search terms? Check out the University of South Florida Free Association Norms database (luckily, searching exactly those words will get you right there). Basically, the database gives you a stimulus word, like *dinner*, and then shows what words people tend to associate with it, like *supper* and *eat*. It also shows how strongly we associate these words with the stimulus. For example, if we go back to *Family Feud*, if the stimulus word is *dinner*, then *supper* would be a bit higher on the list than *eat*. Try it yourself. Head to the database and play around. How closely do your associations match the associations of people polled for this sophisticated version of *Family Feud*?

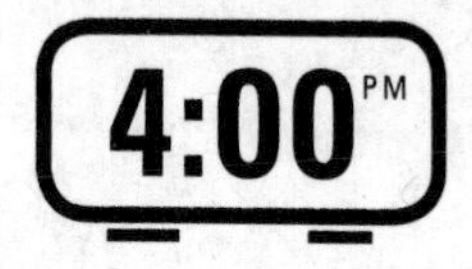


## *HOW TO BE AN INNOVATOR*

It's 4:00 p.m., and unless you're slammed, there's a good chance you've handled your crises for the day and are settling into the afternoon's downhill coast. Here's another way to look at it: you've done most of the things that you know you have to do every day—you've reached the end of the tasks assigned by your expertise—and it's now the time of day when you can choose to stagnate or push forward. There's even another way to look at it: with your crises handled, you've finally got some breathing room for creativity and maybe even innovation. Here's how to do it.

First, studies primarily by K. Anders Ericsson at Florida State University show there's no magic in becoming an expert—it takes about ten thousand hours of practice to develop expertise at just about anything, including music, sports, chess, and highly skilled professions. If you break it down, that's 250 weeks of practicing for forty hours a week, or just under five years of full-time practice. If you can only find four hours instead of eight to practice every weekday, spread the total time needed over ten years of part-time practice. If you want to play second violin in the Pittsburgh Symphony Orchestra, make a living as a country club golf pro, win more games of chess in Central Park than you lose, or manage the demands of a highly skilled job, start practicing today.

But what does it take to go even further? What if you

want to be an *artist* rather than an *artisan*? What if you want to be Steve Jobs instead of a cog in the machinery of an IT department? Or a PGA Tour golfer rather than a country club pro? What sets apart a composer of music for car commercials from a composer who wins a Pulitzer? Or a scientist who takes the next step from a scientist who takes a great leap? Ten thousand hours of practice will make you an expert, but you'll need something more if you want to be an innovator.

Psychologist Dean Keith Simonton, a leading creativity researcher, shows that this something may be equal parts logic, zeitgeist, and luck. For example, Simonton points to the work of computer programs like one called BACON (after Sir Francis Bacon, not after the wickedly delicious pork product), which uses logic alone to come up with "creative" ideas like Ohm's law and Kepler's third law of planetary motion. Instead of springing primordial from the ether, your innovative genius may spring from the ability to deduce new things logically from old information. The flip side is that "great scientists are precisely those who have the ability to dispense with logic," Simonton writes. In other words, making deductions that turn out to be logical in hindsight may require thinking in a way that wouldn't *seem* logical to most people at the outset.

Then there's the perspective of zeitgeist, which imagines creative genius as the inevitable product of the times: only the late 1960s could have created Jimi Hendrix, and the fact that so many Nobel Prizes are shared between concurrent, independent discoverers shows that genius ideas may be floating in the cultural ether waiting to be grabbed.

Finally, there's blind luck. Creative people stand on the shoulders of other creative people, but the only people we recognize as "innovators" are those who happen to be standing there with the innovation or discovery in hand when it finally comes to fruition. Did you think that James Watt invented the steam engine in 1781? That's because Thomas Newcomen had made significant inroads in 1712, and Thomas Savery patented what he called a "steam pump" in 1698.

These three things, logic, zeitgeist, and luck, can help you innovate—say, by working to make logical conclusions from what has come before, or by living at the center of the zeitgeist (e.g., by moving to San Mateo if you make software or to Portland, Oregon, if you make beer).

There's one more thing you gotta do if you want to innovate, and we hinted at it in the beginning: you have to avoid the trap of your own expertise and maintain flexibility even once your experience shows you the "best" ways to do things. It's not easy. For example, when an obvious "best" chess move exists, expert players may miss innovative options, and when medical symptoms present in classic clusters, doctors may miss a non-routine diagnosis. In these cases, it's the very fact of expertise that squishes innovation.

Even if some combination of logic, zeitgeist, and chance is in your favor, innovation still requires constantly asking what's next, even when you think you already know the answer.

## Creativity + Practical Intelligence = Innovation

Creativity and innovation aren't the same thing. Creativity is something that you can do in the privacy of your own basement; innovation is what happens when creativity connects with an audience. And studies show an ingredient that makes a major difference between the two: in addition to creativity, innovation requires practical intelligence. Only by understanding how to reach people can creativity become innovation.

PART 3

# THE EVENING

## *HEALTH, FAMILY, LOVE, AND LEISURE*

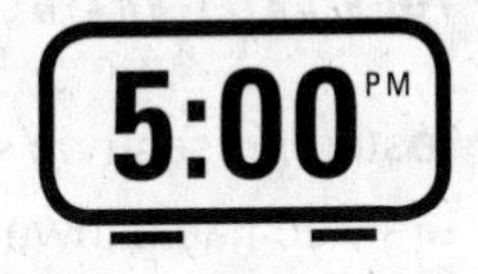


## *DOES YOUR BRAIN LIKE ONE DRINK A DAY?*

It's five o'clock somewhere, and that somewhere is here! But short of becoming a professional boxer or a depth-record skin diver, drinking too much alcohol over time is about the worst thing you can do to your brain. With heavy alcohol use, short-term memory, long-term memory, and motor skills all decline. Even after years of abstinence, alcoholics fail to recover long-term memory. But what about those moderate drinkers we've all heard about? Does one drink a day—preferably red wine—really increase cognitive performance? Does your brain appreciate a gentle ethanol wash?

Let's go back to the source. You don't get more authoritative than a 1988 study published in the *New England Journal of Medicine* that explored coronary disease and stroke in a population of 87,526 female nurses. The article's authors write, "Among middle-aged women, moderate alcohol consumption decreases the risks of coronary heart disease and ischemic stroke." It's the study that launched a thousand ships! Drinking is good for you!

But beyond the beneficial effect of three to seven drinks per week on the circulatory system, things get just a bit fuzzier. Let's look at these same nurses, this time in 1999, when 12,480 of them were between ages seventy and eighty-one. Again, "moderate drinkers had better mean cognitive scores than nondrinkers." When they tested nurses two years later, these older moderate drinkers also had a lower chance of cognitive decline.

Now, the reason for better brains in moderate drinkers is a little tricky to pin down. First, there's the cardiovascular benefit: a bit like aspirin, a little bit of alcohol administered in small quantities over time may thin the blood in a way that helps it pass more easily through the cardiopulmonary system, bringing a bit more oxygen and nutrients to your brain.

Then there's the idea that moderate drinking may do good things for personality and sociability. It's a fact: moderate drinkers are more likely to be social than their heavy-drinking or teetotaling peers. And just like boxing and deep-sea diving will squish brain cells like grapes in a wine barrel, being social is one of the most powerful ways to make brain cells happy. Something similar is true of mood. Since humankind's first experience with fermented juices, we've known that moderate alcohol consumption makes people just a bit euphoric. In modern society, moderate drinking has been shown to reduce stress, cut the chance of clinical depression, and create "overall affective expression, happiness, euphoria, conviviality and pleasant and carefree feelings," according to a review in the journal *Drug and Alcohol Dependence*. In short, moderate alcohol consumption makes you a better person. And there are tons of brain benefits associated with a slightly better mood, from physiological protection against chronic stress to, again, a higher chance of sallying forth more frequently into the social world.

Can you get the cognitive benefits of moderate alcohol consumption without the drinking? The answer is maybe or even probably. You could be social without alcohol, and you could be in a good mood without alcohol. But when you look across a population, it's the people who drink a little—but not a lot!—that sink naturally into this rhythm of good vibes.

**Brain Hack**

**In this headline that drinking is good, keep the word *moderate* firmly in mind. Beyond one drink per day, the health benefits of alcohol take a plunge off a steep cliff. If you fear that cliff, finding a way to relax, enjoy, and be social *without* the drink may be a much better choice.**

## *PUMP UP YOUR BRAIN!*

It's time to pump you up, baby! Don't you just love it? Don't you just love the gym? You may not like stopping by the gym on your way home from work, but your brain certainly does. Just think about the boost in positive self-perception that comes from deliberate suffering, much the way historical fanatics might have worn hair shirts or walked down the road whipping themselves repeatedly with knotted ropes (a practice that informs many exercises in the modern discipline of CrossFit). Okay, the silly austerity of self-imposed masochism isn't the *only* reason to work out. It's also good for the physical structure of your brain.

For decades we've seen the brain-training effects of a physically enriching environment on lab animals: mice with the opportunity to exercise learn faster than mice in the proverbial six-by-eight cell with a cup of water and a crust of dry bread. And it's not just mice. In fact, a huge combination of studies that looked at the relationship between physical exercise and brain skills in kids ages four to eighteen found improved scores on perceptual skills, intelligence quotient, achievement, verbal tests, mathematics tests, memory, developmental level, and academic readiness for kids who exercised. At the other end of the age spectrum, exercise protects older adults from many forms of cognitive decline associated with aging, potentially even pushing back the onset of Alzheimer's disease.

But why?

One reason is the chemicals released in your brain during exercise, like brain-derived neurotrophic factor or BDNF. Don't look for it in the list of ingredients on that jar of macho muscle supplement. Instead, BDNF is a protein made by the BDNF gene that helps the brain change itself to meet the needs of the environment. The brain's ability to adapt is called *neuroplasticity,* and it depends in large part on BDNF. Researchers see BDNF levels go up in exercising humans and mice, and afterward they can see improvements compared to sedentary humans and mice in things like running mazes (that's for the mice, not the humans). When researchers blocked the activity of BDNF in mouse brains, they showed no improvement in maze running after exercise—it was the chemical induced by exercise that increased the rodents' ability to learn about their surroundings.

These chemicals released during exercise also affect mood. For example, you'll read about dopamine, serotonin, and norepinephrine in this book's entry about love. Well, guess what? These same chemicals are secreted during exercise. Dopamine is physically pleasurable, serotonin creates the feeling of awake energy, and norepinephrine increases alertness and arousal. Combined, these chemicals perk you up. There's a lot to be said for taking a nap when you're tired, but there's also a lot to be said for its opposite—when you're feeling physically or mentally tired, cardiovascular exercise can transform the chemical landscape of your brain into one of focused energy.

Exercise also helps oxygen and nutrients move through your body, and part of your body is your brain. The Greeks said *mens sana in corpore sano*, meaning "a sound mind in a sound body," and it's increasingly clear that the link between the two is the cardiovascular system. When you train your heart and lungs, they are better able to deliver oxygen and nutrients to your brain—studies have shown up to a 40 percent difference in gray matter volume between people who exercise and those who don't in areas including the cingulate cortex, the prefrontal cortex, sections of the dorsal anterior cingulate cortex, the supplementary motor area, and the middle frontal gyrus. You won't be tested on the names of these structures—the point is that exercise creates cardiovascular health, and cardiovascular health literally creates larger brains. This is true to the point that when people are *prescribed* exercise, researchers can see gains in brain volumes. For example, a 2011 study published in the *Proceedings of the National Academy of Sciences* with the optimistic title "Exercise Training Increases Size of Hippocampus and Improves Memory" shows just that: in a sample of older adults, the hippocampi of people who exercised grew by about 2 percent over the course of the year, while the hippocampi of people who didn't exercise shrank by about 1 percent. The memory outcome makes sense: your hippocampus helps you create new memories.

Despite the demands of most indoor jobs in the Internet age, humans are not meant to sit and think. (Take a look at carpal tunnel statistics and you'll see we're not meant to sit and type, either.) In fact, the traditional idea of sitting and thinking may be just about the worst way to bring brainpower to bear on tricky problems. If you want your brain to function at its best, give it what it wants: exercise.

### You Gotta Love It

A host of studies shows why we don't exercise: we think it's going to suck. And the long-term reward of better health can't touch the short-term dread that exercise will be difficult and awful. If you want long-term health for your body and your brain, it's time to find some kind of exercise that you don't dread. Be it *Sweatin' to the Oldies* or walking the dog, unless your brain works differently than the brains of most people, the only way you'll stick with your exercise routine is if you like it.

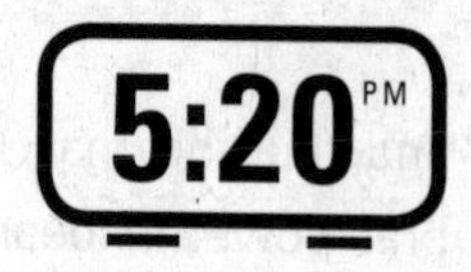


## YOUR BRAIN VS. THE GROCERY STORE

You have a list. You have a strategy. And the goal of the grocery store is to make you deviate from this strategy so that you buy things you don't really need. There are pitches to match every persona: flashy plastic toys hung in the canned food aisle to tempt the hands of toddlers; neon-colored drinks with lightning bolt symbols that promise to reshape middle-aged middle managers into Greek gods and goddesses; produce glistening with dew and right next to it the organic alternative that asks, *Isn't your family's health worth another 34¢ per pound?* Once you've run this gauntlet, you're finally ready to congratulate yourself at the checkout stand. And self-congratulation is right there at eye and hand level in the form of candies and salacious magazines. You bought organic bananas and didn't buy the baker's dozen carton of glazed donuts—don't you deserve a Milky Way?

Like the rides at Disneyland that spit you into shops that merchandise the ride's characters, or a Vegas casino with inexpensive slots by the doors, the grocery store is a system designed to shake you upside down until every last bill and coin and plastic-mediated promise to pay has spilled from your body. How successful you are at remaining solvent in the face of this great shellacking depends on your brain.

Researchers can look inside your brain to see buying

decisions being made in real time. There are two general structures at play: your amygdala and your prefrontal cortex. Your amygdala lusts after that Milky Way, and your prefrontal cortex wonders if it's really a good idea. When researchers use fMRI to look at activation in these brain regions, the balance can help to predict whether you buy or abstain. The fascinating and potentially evil field of neuromarketing also looks at how products can be packaged and priced in ways that make you reach for your pocketbook. Really: if a product's packaging promises something that speaks to your amygdala, you want it, and if the price is just low enough, your amygdala has the power it needs to overrun your prefrontal cortex. In lab studies (but not yet in wide use), marketing techniques that balance amygdala activation with prefrontal activation can help companies price their products to maximize profit without losing the sale.

This scale on which you weigh "want" and "should" in your brain is also the genius of the 1980s Miller Lite advertising slogan "Tastes great, less filling." Your amygdala anticipates the taste, and your prefrontal cortex loves the idea that this lite beer is a healthy option.

But this want/should scale is only the backdrop against which you make buying decisions, after which everything from mood to hunger to personality to money to stress presses down with its thumb on one side of the scale or the other. For example, if you have less time available to shop, fewer items from your list and more unplanned items are likely to end up in your basket. That makes sense:

decisions made while stressed for time aren't likely to be as thoughtful and measured as decisions made with your rational brain more firmly in the driver's seat. But check out this unfortunate corollary, which works through a similar channel: people who have less money may be more prone to making impulsive purchases in the checkout line. It comes down to something social psychologist Roy Baumeister calls *ego depletion* (also known as decision fatigue). It seems that after making many difficult decisions we break down and take the easy choice later on. For example, when researchers have offered snack choices after laboratory tests, they see that when a test is difficult, test takers choose donuts over fruit—their ego is so depleted from the test, they fail to fight off temptations they might refuse in a more ego-full state. (You can only fight snack food for so long!)

Grocery store delicacies aren't the only things that create ego depletion. For example, Princeton economist Dean Spears offered villagers in India the option to buy bars of soap, deeply discounted but still a stretch for many. Afterward, Spears asked villagers to hold their hands in a bucket of ice water—a task that is a common proxy for willpower. With their ego depleted by the difficult soap-buying decision, the villagers held their hands in the ice water for a shorter time. Here's the fascinating part: the reduction in willpower as measured by the ice bucket task was exactly proportionate to villagers' income. When the decision to buy soap was more stressful, weighed perhaps against the decision to buy other necessities, it caused more ego depletion and lowered villagers' willpower for subsequent tasks. The less money a person had, the less willpower they had after making financial decisions.

Let's take it back to the grocery store. Food shopping in general requires decisions, but food shopping on a budget requires *difficult* decisions. Now imagine two very different shoppers as they reach the cash register: one has spent the shopping experience moving from one fairly easy decision to the next and knows that his or her debit card is going to work on the first try, while the second shopper has just spent forty-five minutes weighing the need for food items against the need for car repairs and is trying to keep in mind the sequence of credit cards that might work. And there's a Milky Way bar, displayed so that both shoppers can't help but feel the temptation. Keeping in mind ego depletion and the want/should teeter-totter in your brain, who do you think is going to throw the candy bar on the checkout counter?

## Led by Your Nose

You know about music and lighting design, but a new field of sneaky marketing deals with a store's smell-scape. For example, when researchers odorized areas of a casino for a weekend, slot machine use went up 45 percent compared to non-odorized areas. And a fascinating study in *Marketing Review* shows that store brands can hijack the goodwill of scents provided by premium brands—smelling a posh hand cream makes you evaluate a knockoff brand more favorably. Some department stores distribute scents through the air-conditioning systems, and others use scent stations to customize the scents of different departments. Next time you're shopping, in addition to being aware of the music and intentionally placed impulse buys, take a whiff: do you think that smell is accidental?

## *WHY KIDS PUNCH EACH OTHER IN THE BACKSEATS OF CARS*

You picked them up from school, dropped them off at an activity, picked them up again, and are headed home. The whole time, they've been punching each other in the backseat for no apparent reason. The first thing to check is if there is an unequal distribution of digital technology: nothing creates backseat strife like the desperate unfairness of one iPad split among more than one child. That said, there's a complex formula for fairness. For example, if a child plays on a digital device while waiting for another child's activity to end, does that mean that when the activity is over, the previously busy child should get the screen? If reading a book earns an older child screen time but a younger child isn't yet able to read, does playing an educational iPad game earn the younger child recreational iPad time?

These are important calculations with the power to create either peace and harmony or rancor and overall transportation misery. And these calculations aren't limited to your kids. How do you divide chores with your partner? What's fair? Does picking up dog waste in the backyard once a week balance doing the dishes every night? Does a little skilled labor like rehanging a ceiling fixture balance a lot of more general labor like folding laundry? If one of you takes the kids to a birthday party at an indoor arcade so that the other has time to work, does the birthday party parent earn the right to free time after the party while the parent who stayed home to work watches the kids? Does the calculation change if the birthday party was somewhere slightly less hellish than an indoor arcade? Does it change the measure of fairness if there were adult beverages at the kids' birthday party?

The idea of fairness is subjective and open to negotiation. But when the scale of fairness tips, bad feelings are the result.

This is not a new problem. Almost all religions and philosophies have thoughts on fairness. For example, the Edda, a collection of thirteenth-century Icelandic verses that forms much of the modern understanding of medieval Norse mythology, says, "Man ought to be a friend to his friend and repay gift with gift. People should meet smiles with smiles and lies with treachery." You can almost see some poor dude with a big red beard and horned helmet sitting in the front of the longboat scratching these words on vellum sheets as his kids try to reach far enough across the back bench to whack each other with small stone hammers, the Viking equivalent of a fast-food toy.

More recently, the idea of fairness has jumped from the realms of philosophy, religion, and government into the research labs of economics. Take the famous ultimatum game. In its classic form, two people are splitting a pile of money. The first person proposes a split, and if the second accepts it, they both get paid. But if the second person rejects the proposed split, neither gets anything. Imagine both players are rational. If player A suggests splitting a pile of pennies ninety-nine to one, player B realizes that one penny is better than none and so accepts the proposal. But that's not at all what happens in real life. In real life, player B tends to tell player A where player A can stick that single penny—it's not rational, but it *is* fair. In fact, a long line of research shows that proposed splits that are more than 20 percent unequal tend to be rejected.

Why would you reject a split that gave the other person sixty-one and you thirty-nine? I mean, you still get thirty-nine, right? One answer is that by fighting for fairness, you promote the continuation of a human society in which neighbors help neighbors, we all respect traffic signals, and no one litters in public parks. When people play more than one round of these economic games, researchers see that everyone is more fair. It makes sense: if you expect an eye for an eye, you probably won't act in a way that removes that first eye.

You can see this in another famous economic game, the prisoner's dilemma. In the PD, two crooks are caught and imprisoned in separate cells. Now they have to decide independently whether or not to rat out the other person. If both refuse to talk, both serve one year on a reduced charge. If one squeals and the other stays quiet, the one who cut a deal goes free and the other serves three years. If they both squeal, they both serve two years.

Now, no matter what your partner in crime does, you're better off squealing: if your partner stays quiet, you get off instead of serving one year, and if your partner squeals, you serve two years as opposed to three. But if you could trust that the other crook would go against his personal self-interest and keep his dirty mouth shut, you could do the same and minimize the overall time served—you would each serve one year for a total of two years served, as opposed to some combination of jail time that equals a total of four years.

The prisoner's dilemma and the ultimatum game seem like artificial creations of people who spend too much time with numbers and not enough time with people . . . until you look through the lens of these games out into the world. The arms race of the Cold War was a great prisoner's dilemma: both the United States and the Soviet Union were better off building weapons no matter what the other country's decision was (but if both had stopped, both could have benefited). Doping in pro cycling and baseball is another example: if no one used human growth hormone to grow to the size of a small island nation, then no one would have an advantage; if everyone used HGH, the competitive advantage would be erased too, and everyone would end up with nerve pain, swelling, and carpal tunnel syndrome. But if some people use while others do not, the users win trophies and endorsement deals while the non-users end up adjusting training wheels in the back room of the local bike shop. Like the prisoner's dilemma, stopping doping in sports requires athletes to trust that everyone will act fairly.

Another solution is the human experience of fairness itself. When someone gets away with cheating, a sibling sneaks more iPad time, or your partner watches bad reality television while you do the dishes, you know you've been wronged. And you are tempted to exact revenge, even if it means giving up personal gain to get back at the people who have acted unfairly against you. In the ultimatum game, your irrational internal experience of fairness makes you reject a nine-to-one split because it's worth giving up a coin to keep the unfair a-hole proposing the split from earning nine. With the idea that what goes around comes around hardwired into our brains at the level of the species, we are pulled toward acting fairly.

The problem is that what is "fair" depends on your point of view. A person offering three coins while keeping seven may say, "Gosh, I'm so kind to be giving that person three coins that he or she wouldn't have otherwise." But the person looking at three coins on the table sees it differently: to that person, the only "fair" split is 50/50. Only when one opinion of "fair" overlaps with another opinion of "fair" is there room for a deal that pleases everybody. And that is one major reason why most ultimatum games end with a 60/40 split—a division that both parties can live with.

This is the purpose of the arguing that goes on in the backseat: to find a solution that meets both parties' definition of fairness. And when there is no middle ground between opinions of fairness—when one kid hogs the iPad or when your partner doesn't lift a finger around the house—the difference is expressed in outrage. That's when the winner is forced to pay a "fairness penalty"—when, throughout our long history as a human species, one child punches the other in the backseat of transportation.

## Hack Your Kids' Self-Centered Fairness

Fairness is in the eye of the beholder. A 2002 study in the *Journal of Applied Psychology* shows this is even more true in individualistic cultures like the United States than in collectivist cultures such as Japan—if you feel like your personal power is more important than the good of the system as a whole, your idea of fairness can be very self-centered. So make the little gremlins in your backseat think collectively. Studies have shown that displaying symbols from a collectivist culture should do it . . . maybe hang Mao's *Little Red Book* from the rearview mirror? Researchers have also primed collectivist thinking by asking subjects to think about the groups they belong to, to brainstorm ways they are like other people, or to list reasons it might be good to "blend in." If you can find ways to help your kids remember that your family is in it together, you may all survive the ride home.

## *HOW PARENTING CHANGES THE BRAIN*

Do you have kids? If so, your brain is different than the brains of people who don't. Most of these changes make you less likely to eat your young this evening after they have punched each other in the backseat of the car on the way home. Take mice, a species in which it is much easier to insert wires into the brain to watch the growth of new neurons than it is in, say, humans. In the days following the birth of his pups, a burst of new neurons grows in a mouse dad's brain—that is, says a study in *Nature Neuroscience*, if he stays in contact with his pups. Smell alone isn't enough to do it, but with physical proximity, the new mouse dad grows neurons in his hippocampus (memory) and olfactory bulb (smell) that help him recognize his pups' individual smell, a skill that remains through life and allows the mouse dad to continue to recognize his pups even once they're grown. The key is the hormone prolactin, the same hormone that switches on lactation in mothers.

Fatherhood also decreases the production of testosterone in human males. A 2011 study in the *Proceedings of the National Academy of Sciences* shows that men with high testosterone are most likely to mate, after which testosterone drops by an average of 30 percent. Here's the very cool part: like a male mouse growing neurons based on the degree of contact he has with his pups, the degree to which a man's testosterone drops when he becomes a father is directly related to how much time he spends with his child. You might think of testosterone as a "body" thing, and it certainly is, contributing to the formation of muscle, but it's also a brain thing, and a drop in its levels means that Dad becomes less aggressive and more nurturing—in short, he becomes more dad-like and less dude-like.

There's also a distinct difference between the brains of parents and non-parents in response to the sound of a baby's cry, even if that baby is not their own. When a baby cries, the parental brain shows activation in the amygdala and prefrontal cortex, as if the parent's brain is wired to feel the baby's anxiety (amygdala) and immediately ask what can be done about it (prefrontal cortex). The brains of non-parents? Not so much. To the brain of a non-parent, the sound of a crying baby is no different from any other grating noise.

Really, "parent brain" is just one in a long line of age- and life-stage-related changes in your gray matter. When you were very young, you didn't know anything, and so your brain was wired for super-fast learning so that you could come to know something. In the teenage years, our ancestors had to fight for places in the social hierarchy, which sometimes required risky and aggressive behaviors, so the teenage brain is wired for this kind of flamboyant idiocy. As a young adult you were programmed to search for a mate, and if you were successful, your brain reworks itself to prioritize the needs of the children you're parenting.

Tonight it's finally time to stop fighting your brain. If you're a parent, realize that your brain will never be the same as it was before you had kids. But that's not necessarily a bad thing. Rather, it's the result of your brain's super-cool ability to adapt itself to your needs. If you're a parent, you need a parent's brain. If you're not, your brain can continue to hear a baby's cry as no different than fingernails on a blackboard.

### Fake It Till You Make It

The moral of all this parental brain stuff is that actions lead to development: the more you act parental or non-parental, the more your brain is likely to transform itself to follow suit. The amount of time moms and dads spend with their child creates mom-like and dad-like brain states. Nurture your kid and chances are your brain will follow.

## *LOVE IS A HEADY CHEMICAL COCKTAIL*

If you want to know what's going on in your brain this evening when you experience love, you first have to define your terms. Do you mean *falling* in love or *being* in love? Both are like a drug. If you've ever felt euphoric and dizzy after a promising date, you know that falling in love mimics the body's response to a recreational drug. However, not to hit Cupid with a bag of poo or anything, but *being* in love is very different—more like waking up two weeks later in a pile of Cheetos bags with the dawning comprehension of addiction.

That's because falling in love depends in part on good old dopamine—the brain's reward chemical that is mimicked by opioids like heroin. Love, like heroin, creates dopamine release in the brain. Not only does this dopamine make you literally a bit dopy, blocking pain and inducing pleasure, but the pleasure of dopamine release makes you do things that lead to more pleasure. The dopamine in your brain after a good first date helps to ensure there is a second date.

Then you also get some oxytocin. No, this isn't the widely abused prescription drug OxyContin or the equally abused oxycodone, but a naturally occurring neurotransmitter that facilitates the process of human bonding. Actually, oxytocin doesn't only create human bonding; it's been shown to create the mom-pup bond in lab mice. A nasal spray of oxytocin makes people in experiments more cooperative. And, just as it makes you bond with people you see as friends, allies or lovers, oxytocin also makes you exclude people you see as outsiders—meaning that a squirt of oxytocin while falling in love helps you grow eyes for your loved one while blinkering you to other options.

During sex, oxytocin is joined by vasopressin, a hormone involved in the body's regulation of water retention and vasodilation. Research with prairie voles shows that vasopressin might be what creates the bonding experience of sex—with vasopressin, voles bond after mating; with vasopressin blocked, voles leave it at a one-night stand. Here's a cool part: when researchers introduced vasopressin to pairs of prairie voles, they bonded even *without* mating. At least in prairie voles, it's a chemical cocktail and not a whole lot more that creates the bonded experience of love.

And gosh darn if we can't see the same thing in humans—the better your brain is at using oxytocin and vasopressin, the more likely you are to be in a bonded relationship. Really: your brain cells have receptors that are designed to trap these hormones, and there's some evidence that people with less oxytocin and vasopressin receptors are more likely to divorce and more likely to have lower relationship satisfaction.

But that's not the end of it. Like a molly pusher who cuts MDM with all sorts of chemicals meant to keep pests out of your garage, love mixes even more chemicals into your poor, besotted brain. One of these is norepinephrine, which makes your heart go pitter-pat. Norepinephrine, released during the elation of love, is part of the body's fight-or-flight response, and it increases your ability to skip through fields of tulips with your new mate.

Finally, there's serotonin. "Love lowers serotonin levels, which is common in people with obsessive-compulsive disorders," says Mary Lynn, DO, codirector of the Loyola Sexual Wellness Clinic. "This may explain why we concentrate on little other than our partner during the early stages of a relationship."

So you've got dopamine, which makes you stoned; oxytocin, which makes you dependent; norepinephrine, which makes you tweak; and serotonin, which makes you obsess. Over time, you become unable to exist without this stew of neurotransmitters. The longer you are in a relationship, the more your brain becomes used to these chemicals, and you reset your baseline in a way that makes you *need* them. The comparison with addiction is so strong that when Rutgers researcher Helen Fisher showed freshmen who had recently broken up with a boyfriend or girlfriend a picture of their ex, she saw the same areas of the brain activated as in drug cravings.

It makes sense. Our brains didn't evolve to feel pleasure in response to drugs; drugs are simply an artificial way to hijack the pleasure system that was designed for another purpose, which is to help you feel love and bonding. To your brain, love is a drug and drugs are love. Put that in your pipe and smoke it.

## Touch, Sunlight, Love

Short of searching Amazon for legitimate dealers of oxytocin nasal spray (don't waste your time) or putting your PhD to use by genetically engineering your way to increased serotonin release, you'll have to come by the brain chemicals of love the old-fashioned way: by experiencing love. If love is hard to come by, physical contact will do it. Even platonic hugging creates oxytocin release, and some researchers believe that people who crave human contact after isolation are experiencing a sort of oxytocin withdrawal. And a 2007 article in the *Journal of Psychiatry and Neuroscience* recommends bright light to create a serotonin release similar to being in love. You thought "basking in the love" of your partner was only a metaphor? Turns out to be a little more literal than that. If you're starved for love and human contact, you can get a hint of it by getting out into the sunlight.

## *HOW TO KNOW IF YOUR RELATIONSHIP WILL LAST*

Now that you know what's kicking inside your brain while you're in love, the obvious question is, will it last? John Gottman has the answer. The famed relationship researcher and professor emeritus of psychology at the University of Washington spent a career unpacking words, actions, reactions, and emotions inside a relationship that can predict success or failure.

For example, he videotaped married subjects talking to each other while measuring skin temperature and heart rate, asking them to press levers to indicate the strength of their feelings while talking, and then having trained graduate students code the couple's facial expressions while they were talking. It turned out there were four things that he consistently saw in couples who later went on to divorce: criticism, defensiveness, stonewalling, and contempt. You probably know the general meanings of these things, but it's worth zooming in to see what exactly it is within these things that Gottman calls the "Four Horsemen of the Apocalypse" that makes relationships fall apart.

Sure, there are things your partner does that you dislike—you hate it when he forgets to push the "dark brew" button on the coffeemaker and the morning joe comes out all watery; you hate it when she downplays her computer programming experience at parties. And you want to tell this person you love about these things you dislike; you want to *criticize*. Gottman says go right ahead. Whether or not it's realistic to expect your partner to change these things he or she does that you dislike, it's completely unrealistic to expect you to bite your tongue when you see things in the relationship you wish were otherwise. It's not criticism of a partner's *actions* that Gottman lists as one of the Four Horsemen, but criticizing your partner's *character*. When a complaint about the way your partner does something transitions into criticism about the part of your partner's character or personality that makes him or her act this way, it's a marker of way-bad relationship juju. Think about it: do you say "I hate it when you leave the toilet lid up" or "What kind of inconsiderate a-hole leaves the toilet seat up!"?

Defensiveness is, as it sounds, the attempt to deflect attack. The form of defensiveness that really kills relationships is more like a blame mirror—instead of simply dodging an attack, you turn it back on your partner. It's also being quick to see attack where it might not even exist. This kind of defense isn't avoidance or constructive debate. It's more like *offense,* fighting fire with fire even if it means preemptively throwing fire at a partner who hasn't yet lit his or her torch.

Stonewalling is that faraway look your partner gets while you're criticizing. It's disengaging from an interaction without bringing it to a conclusion, or refusing to engage in difficult exchanges in the first place. If a partner is stonewalling, it's as if he or she has already checked out.

Then there's contempt. Contempt is a bit like criticism,

but with mocking disrespect. Contempt is not only wishing that your partner was different, but devaluing how he or she is. To Gottman, the *puh-leeze* accompanied by an eye roll is the most destructive of the Four Horsemen, indicative of long-term resentment that has been simmering below the surface.

These sound like the breezy observations of a self-help blog post until you look at where they came from. Starting in 1989, Gottman recruited 130 newlywed couples from the Seattle area into a longitudinal study: all were childless and had been married fewer than six months. He had them fill out questionnaires and then complete a taped interview during which they discussed an ongoing problem of their choice. Then Gottman followed these couples for six years. During that time, there were seventeen divorces. What had these couples done differently than the 113 still-married couples during their interviews in the Gottman Lab?

That's where the Four Horsemen came from—the couples who later went on to divorce were much more likely than others to criticize character, act defensively, stonewall, or fix their partner in the spotlight of contempt. Gottman's model predicts divorce with 83 percent accuracy, and it predicts marital satisfaction with 80 percent accuracy. This means that if you see the Four Horsemen in your relationship, there's still a chance you'll be just fine . . . but it might be time to look a little more closely at how to get off the horse of relationship badness before it's too late.

## "Do You Like Horror Movies?"

To know things about a couple's compatibility, you can watch a lot of couples or, in the Internet age, you can watch a lot of numbers. That's what Harvard mathematicians and founders of the online dating website OK Cupid did to discover the three questions on their compatibility questionnaire that best predict which couples will end up working out. Here they are: "Do you like horror movies?" and "Have you ever traveled around the country alone?" and "Wouldn't it be fun to chuck it all and go live on a sailboat?" It may seem silly, but more so than thousands of other survey questions, these three in combination do a darn good job of predicting long-term compatibility.

## *SHOULD YOU ARGUE WITH YOUR SPOUSE WHEN YOU'RE ANGRY? HOW ABOUT WHEN YOU'RE SAD?*

There it is again, that little devil of anger sitting on your shoulder, jabbing you in the neck with that damn pitchfork. You know that little guy is going to keep prodding you until you fight with your significant other. We've all heard that it's unwise to argue when you're angry because you'll end up shouting something you don't really believe and can't take back, and generally making a mess of the world as you know it. But due to the lovely inner workings of your brain, not only are you more likely to lose control of your behaviors when you're angry, but you may also lose control of your beliefs. How you interpret your partner's intentions, character, and genetic similarity to slime mold depends on your mood—and you'd better bet that when you're mad, you're primed to believe the worst.

It's like this: Imagine you're driving through a crowded grocery store parking lot when a man on a cell phone walks out in front of you. While still holding the phone to his ear, he holds up a hand toward you, palm out. What does he mean by this gesture? And what is the appropriate response? Should you show him your palm in return, or should you show him the reverse side of your longest finger? How you interpret the man's gesture—an apology, a thank-you, or a command—depends in large part on your own mood. If you're in a bad mood, you're more likely to see this man's palm as an insistence on his entitlement to walk through traffic while yakking on his phone; if you're in a good mood, you're more likely to give him the benefit of the doubt and see his gesture as an apology for the inconvenience and a thank-you for not turning him into a pavement pancake.

This is due to something called the "mood congruency" effect, and it means that when you feel bad and nasty, you think other people feel bad and nasty. That makes sense, but the thing is, not all flavors of "bad and nasty" look the same in your brain. Anger lives in your amygdala—it's the "lizard brain" flavor of bad mood that cranks your pulse, blood pressure, and secretion of epinephrine. But sadness lives in the hippocampus—it's a more cognitive experience of bad mood that draws on memory and your interpretation of experiences, without necessarily affecting your body. Do angry people interpret social cues differently than sad people? Let's ride along with Galen Bodenhausen, professor of psychology at Northwestern University, and take a look.

First he asked college students to "vividly recall an episode that had made them feel very angry, and describe in detail how the event occurred." Others did the same thing with a sad memory. A third group was allowed to keep whatever mood they brought with them to the lab. Then he had students imagine they were sitting on a peer review panel judging cases of student misconduct, one involving cheating and one involving assault. In half of these cases, the fictional defendant was given an obviously Hispanic name. How did these sad, mad, or neutral students judge their Hispanic or race-neutral peers? Unfortunately, as you may have guessed, when college students were angry, but not when they were sad or neutral, they were much more likely to see guilt in peers with Hispanic names, but no more likely to see guilt in cases that included an accused person with a race-neutral name.

To Bodenhausen, this is evidence of "heuristic information processing"—when you're angry, instead of using your rational brain, you go with your gut, and in this case students' guts included stereotype. In other words, anger made students lose their minds.

Then he did the same thing with persuasion. Bodenhausen again made college students happy, sad, or neutral and then had them read an essay arguing to raise the legal driving age from sixteen to eighteen. Half the college students were told that the essay had been written by "a group of transportation policy experts from Princeton University," and half were told it had been written by "a group of students at Sinclair Community College in New Jersey." How persuasive did college students find these arguments? It turned out that sad students left their rational brain in charge and formed their opinions largely on the content of the written argument; for angry students, the source of the argument trumped the content. Bodenhausen found the same thing when he varied the sources' trustworthiness rather than expertise: when information came from a biased source, angry students let their distrust of the source overwhelm the information, whereas sad and neutral students distrusted the source but still based their opinions on the content of the argument.

What this means is that anger blunts your ability to be rational. You probably already knew that. But anger blunts rationality in a very interesting way, as if it opens a direct channel of communication to your biases and instincts and heuristics—to all the beliefs and decision-making rules of thumb that are your fallbacks when not overwritten by your conscious mind. When you're angry, you judge the person and not the content. You're also persuaded by the person and not by the content.

If you argue with your partner while you're angry, your opinion of that person as idiotic and untrustworthy trumps anything smart or real or insightful he or she could possibly say. On the other hand, if you're trying to stay focused on your side of a problem even in the face of a charismatic and persuasive partner, your anger may make you more persuadable. You can argue when you're sad and still stay logical. But arguing when you're mad puts your lizard brain in the driver's seat and, too often, your best interests in the trunk.

## *A TINY PIECE OF EVIDENCE THAT SUGGESTS YOU STILL HAVE THE BRAIN OF A CAVEPERSON*

As you choose how you will interact at home this evening with the ones you love, or at the gym, restaurant, or singles bar with ones you would *like* to love, it's good to know that your brain allows you to be so much more than the block-headed tool of reproduction that evolution built. Since at least the Enlightenment or maybe the Industrial Revolution, we are unbound by the laws of evolution, free from the meddlings of Darwin in our brains. Instead of being beholden to our animal instincts, we are rational beings with an evolved purpose that goes beyond the need to procreate (even if we can't exactly articulate what this purpose might be). But are you really? Are you really beyond evolution?

We could strip the northern forests of pulp for the paper that it would take to describe all the studies showing that men and women are still doomed to act according to our evolutionary impulses. For example, take a paper published in *Nature* in 2000. It starts by naming past studies that have shown that tall men have an advantage over short men in academics, health, income, and social status. Really, that should be enough on its own: why should height influence your career success? But the study goes even further, showing that tall men have more children.

Even in this enlightened age in which we live, being tall gives men a basic evolutionary advantage.

As highly evolved as you undoubtedly are, there's still room to grow.

## What to Talk About on a First Date

The behavioral economist Dan Ariely showed that people on first dates go about their conversations all wrong. We tend to speak in generalities, talking about the weather or food—topics that guarantee a vanilla sort of agreement. But according to Ariely, sticking to safe topics like hobbies and jobs makes a middle-of-the-road date, quickly forgotten. When Ariely forced daters to choose from a list of pointed questions like "Have you ever broken someone's heart?" and "How many romantic partners have you had?" both the askers and the answerers rated the conversations as more enjoyable. Settling into topics that guarantee equilibrium is safe. But shaking it up is the path to a good date.

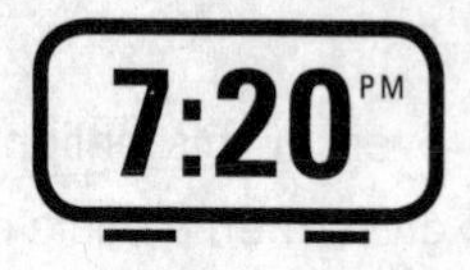


## *HOW TO READ YOUR SIGNIFICANT OTHER'S MIND*

"Hey, honey, what do you think about inviting my mom over for dinner this weekend?"

"Instead of a ski vacation this year, what about a Disney cruise?"

There are politically correct ways to answer these questions—ways that will keep one of you from sleeping on the couch. And then there are the *real* answers—what you are really thinking behind that polished veneer of politeness and compromise. If you want to know what your partner really thinks, you need a window past the machinations of his or her prefrontal cortex—past the mask that we are so good at slapping down on our faces to hide our emotions and intentions.

This window exists in something called a *microexpression.* Like the delay between seeing the pep band's cymbal crash and hearing its roar, there is a short delay between an emotion we feel and our ability to mask it. In these milliseconds, the human face shows its true emotion.

A quick image search online for "Ekman microexpressions" (after the researcher Paul Ekman, who coined the term and has contributed most of the major studies in the field) shows hundreds of good examples of microexpressions—pictures that show faces caught in the act of emotional expression. But in the real world, one secret to spotting these microexpressions is *not blinking.*

Ekman shows that microexpressions last only one-fifteenth to one twenty-fifth of a second, after which people tend to consciously or unconsciously conceal them.

This small window offers important information. One study found that a doctor's ability to recognize his or her patients' microexpressions correlated with patients' evaluations of the doctor's empathy. Another study showed videotape of a speech by George H. W. Bush from which researchers had removed seven smiling microexpressions. Without the speech's friendly subconscious underpinnings, subjects felt more threatened and angry. Another showed that people who can detect microexpressions are better at catching liars—that is, as long as liars are fibbing about something with emotional content, which creates what the researchers call "facial leakage" (meaning that if you're lying, you should try to detach emotion from the lie).

Even subconsciously, microexpressions influence your perceptions of the world, and learning to see them with your conscious mind could give you clues to your significant other's thoughts and feelings that you might not be able to get any other way. The question is, can you learn to notice these microexpressions consciously? Computers certainly can. Citing its use in a lie detection system, Chinese researchers describe an automated system that recognizes and interprets microexpressions with 85.42 percent accuracy, which they write is "higher than the performance of trained human subjects."

But even without a computer's ability to take a snapshot and then progress frame by frame while crunching data at the speed of a silicon chip, you can learn to pull information from microexpressions. Ekman offers a range of (paid) tools to train your ability to spot and use microexpressions. But one thing you can do for free right now is simply work to raise your awareness that microexpressions exist. We miss the information of microexpressions because we don't think to look.

Next time you have an important conversation, be it while evaluating a job candidate, chatting with a friend, or this evening while attempting to glean hidden information from the people you love, watch closely to see if the flash of one expression is replaced by the mask of another. This short window can offer a glimpse into truth, lies, and the real emotions people are feeling.

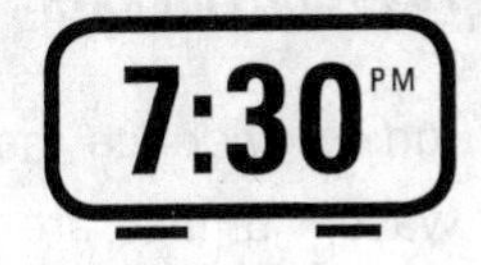


## *HAPPINESS, WELL-BEING, AND TWINKIES*

The U.S. Constitution does not guarantee happiness. In part that's because the whole "life, liberty, and the pursuit of happiness" thing is from the Declaration of Independence and not from the Constitution. But another important reason is that while a government can ensure the right to pursue happiness (or at least the thought makes brilliant propaganda), nothing can guarantee the attainment of this happiness. So as you sit on your couch tonight reading this book and looking back on your day, ask yourself: are you happy?

First ask yourself whether "happiness" is really what you want. If you're sitting there despondent, it sure seems like it is. And even if you're sitting there pretty contented, you gotta admit that you could always be happier, right? I mean, as you're reading this, are you or are you not lying in a hammock on the beach with a margarita at your right elbow? And if you are, how is the temperature? Is the breeze conducive to your happiness? At what magnitude of happiness are you truly *happy*?

Aristippus, a Greek philosopher who lived in the fourth century BC, asked the same question and came up with the simple answer that if you feel good more than you feel bad, you are happy. To him the goal of life was to maximize the times you feel happy while minimizing the times you feel sad. Recently, the field of economics has provided a ton of evidence that this is exactly what we do. In many ways, humans are simple creatures who "maximize utility" by seeking pleasure over pain, to the point that famed Princeton economist Daniel Kahneman recommends measuring your happiness by recording your day in a diary and counting how often and how powerfully happiness overbalances unhappiness.

But might happiness be more than simply feeling good without feeling bad (and maybe the lingering afterglow of smugness)?

That's where the eudaemonists come in. No, eudaemonists are not crackpots who look for hidden meaning in the mathematical ratios of Egyptian pyramids, nor are they people who wear masks and participate in weird orgiastic rituals in Oliver Stone films. Or if they are, it is independent of what makes them a eudaemonist. A eudaemonist is a person who believes, like Aristotle, that there is more to life than the knee jerk of our desires. You can almost hear the cheering of Sam the Eagle from *The Muppet Show* as he describes the benefits of things like strong moral fiber.

From the eudaemonic perspective, happiness can really suck. Take Panama. According to UNICEF statistics, in 2012 Panama had a per capita income of $9,910, only forty-five Internet users per hundred people, and 6.6 percent of the population living below the international poverty rate of $1.25 a day. Compare that with the mighty United States, in which UNICEF stats pegged our 2012 selves at $50,120 per capita, eighty-six Internet users per hundred, and an immeasurably low percent of people living on less than $1.25 a day. *Obviously* with more money and the ability to view things like the Nyan Cat video, people in the United States are happier than people in Panama, right? Not according to the 2013 Gallup-Healthways Index, which polled 133,000 people in 135 countries to come up with a report titled *The State of Global Well-Being*. When you look at the well-being of nations' citizens, Panama was first on the list. The United States was forty-first.

That's because in addition to the in-the-moment experience of happiness, the Gallup poll took into account domains like *purpose* (liking what you do each day and being motivated to achieve your goals), *social* (having supportive relationships and love in your life), *financial* (managing your economic life to reduce stress and increase security), *community* (liking where you live, feeling safe, and having pride in your community), and *physical* (having good health and enough energy to get things done daily).

According to the report, the well-being of people in the United States is bolstered by purpose and by social well-being but dragged down by lack of well-being in community, physical, and financial domains—apparently, despite our economic juggernaut, American people feel as if we can't make ends meet. War-torn Syria and Afghanistan round out the bottom of the list, with only 1 percent of people "thriving" in each. At the top of the list, after Panama come Costa Rica, Denmark, Austria, and Brazil.

Let's imagine that well-being and not happiness is what you want. What can you do about it? One thing you can do to improve your well-being and thus do your small part to improve the well-being of the United States is to find a long-term partner, which boosts the chances of an American scoring among the "thriving" from a dismal 13 percent to a still dismal (but at least suddenly able to look down on someone) 17 percent.

Physical well-being in the United States is also dismal . . . and in an interesting way. Yes, we're talking about obesity, and this is one example of many in which you may have to trade happiness for well-being. Take Twinkies. If you maximize your happiness by sucking back ten or twenty of these delicious cream-filled sponge cakes, you are likely to experience immediate happiness followed by a dip in well-being including but not limited to self-loathing and projectile vomiting. The Twinkie case shows that what feels good isn't always best.

But it's so damn easy to put happiness on one side of a teeter-totter and sadness on the other, and mark the seesaw's tilt as the measure of a happy life! Too bad, sucka: we live in an age of equivocation, and even something as pure-seeming as happiness has been dragged into the muck along with other unassailable goodness like trans fats and the U.S. Congress. And in this equivocal world in which neither money nor education nor under-five mortality is an absolute measure of personal or societal happiness, how you evaluate your life may come down not to what happens to you but to how you frame it. Studies of personality show that "characteristically happy people tend to construe the same life events and encounters more favorably than unhappy people," wrote researchers Richard Ryan and Edward Deci in 2001.

In the immortal words of Bobby McFerrin, if you want to be happy, be happy now. Happiness is in the pursuit, and the choice to pursue it is up to you.

## Hack Your Well-being

Want ways to boost your well-being? The following list may sound banal, but every item has been proven, study after study, to help you feel better about yourself and your life in a meta way: take care of your body, find opportunities to laugh, express emotions rather than repressing them, live according to your values, practice gratitude and forgiveness. (Oh, and shorten your commute.)

## *DON'T DISS THE PLACEBO EFFECT*

As Jack Nicholson famously said in *The Shining*, "It's finally time to take your medicine, you whelp." But which of your fifteen jars of nighttime medication lined up along the ledge above your toothpaste do you really need to take? Maybe all. Or . . . maybe none. That's because placebos are getting stronger! Yes, those sugar pills that are used by doctors and researchers to trick the control groups of clinical trials are working better and better every day! So powerful are these supposedly innocuous and inert placebo pills that it's getting harder to beat them with actual medicines.

This seems great—someday soon you may be able to pop an empty gelcap instead of your antidepressant!—but it also means, as an article in the journal *JAMA: Psychiatry* says, that "increasing placebo response rates contribute to decreasing drug-placebo differences and increasing numbers of failed antipsychotic trials, both of which increase the cost of drug development, delay the clinical availability of new antipsychotic medications, and even precipitate reductions in pharmaceutical company research for psychiatric disorders."

The reason placebos are getting stronger may be due to your brain's increasing confidence in the power of these medications. More so than in the 1960s, we expect to be able to medicate away our problems. We *expect* drugs to work. And so they do. In fact, almost anything that you expect to have an effect will have that effect, from fringe therapies that show astounding health benefits to the "nocebo" effect—men in one study who were told that a common prostate medicine could lead to sexual dysfunction had twice the rate of sexual dysfunction of men who were not told about the possible side effect. We pooh-pooh the power of psychosomatics as "all in your mind," but what is in your mind has the stunning potential to leak out of that bag surrounding your gray matter and into the rest of your body. The power of your mind can make the things you believe about your body come true.

But overconfidence in medicines can also make them work *worse*. When you expect something to work, it does.. But when you expect a pharmaceutical treatment to be a cure or a magic wand that solves all your woes, it may undercut the motivation to do all those other things outside medication that could help you. Hey, if you're taking a magic pill, why should you exercise and eat a healthy diet? In this case, you may take two steps forward with the medicine, but also one step back because of lazy health habits. Back in the day, we distrusted meds and so kept doing everything *else* we could do to be well, making the people on these meds overall healthier and making the effect of the medication seem stronger.

The message for you tonight as you reach for your meds is that there are amazing products on the market that can help your body and your brain heal and maintain itself in amazing ways. But don't forget about all those *other* things you could and probably should be doing. Believing in a pill will help it do its work, whether it's the right pill or not. But believing that a pill is all you need can link the extent of your wellness to an imperfect chemical.

## Crystal Healing

It's time to stop dissing crystal healing and the power of magnet watches. What you believe will work is very likely to work. Do you really care if you feel better due to the power of the crystal or due to the power of your mind? This week, find something you believe in and give it a try. The subjective experience that your brain creates for your body is what matters. (And not the odd looks you get while wearing a tinfoil helmet.)

### *TECHNOLOGY IS EATING YOUR BRAIN*

Admit it: is this book your second source of media consumption right now? What's the first? Is it a smartphone? A tablet? One of those old-fashioned inventions, the color TV? If so, your brain is being eaten alive as surely as if you were a nameless supporting character in a 1970s zombie flick.

Let's go back to that color TV. In 2011 an article in the journal *Dreaming* described an interesting feature of dreams from 2,077 people surveyed in 1993 and another 1,328 surveyed in 2009. The feature was whether people dreamed in color. Overall, younger people dreamed in color—80 percent compared to only about 20 percent by age sixty. That's great: it seems as if color dreaming is a skill of young brains.

But here's the interesting part: color dreaming increased from 1993 to 2009 *only for people in their twenties, thirties, and forties*. For some reason, more young color-dreamers were being pumped into the system. To discover why, you have to look thirty years earlier. From 1964 to 1978—the formative years of this new crop of color-dreamers—the percentage of households with a color television grew from 3.1 percent to 78 percent. The researchers of the article in *Dreaming* blame color television and not young age for the ability to dream in color. It's unclear what percentage of these color dreams were specifically related to *Jaws* and *Children of the Corn.* But what is clear is that technology has the very real potential to change your brain—dude, color TV changes the complexion of your *dreams.*

Technology is also robbing us of sleep. When Thomas Edison invented the lightbulb, he forever unbound humanity from the rhythm of daylight and night. Your cognitive mind can thumb its nose at circadian rhythm, but your lizard brain didn't get the memo—it still wants to go to sleep when it's dark and be awake when it's light. See, it's supposed to get dimmer in the evening as we wind down toward bedtime. When you read on a tablet device, it doesn't have to (and when you play that adrenaline-inducing app instead of reading, your body doesn't get dimmer either). The chemical melatonin is the substance that makes you sleep. When the world gets dim, your brain produces melatonin and you get drowsy. But when the world stays light, no matter the time of day, your eyes tell your brain to produce less melatonin, making it harder to get to sleep. In an article for CNN, Mariana Figueiro, director of the Lighting Research Center at Rensselaer Polytechnic Institute, recommends wearing orange sunglasses in bed while reading on an iPad, as the colored lenses filter the blue light to which our eyes are especially sensitive. (We'll give that one a maybe.)

Even outside the memory-crushing effects of sleep deprivation, technology punks your ability to hold on to facts. In 2007 Ian Robertson, director of the Trinity College Institute of Neuroscience in Dublin, asked three thousand people to remember basic personal information—their phone number, their home address, a relative's birthday. Now, we all know that older people have a tough time remembering things, so cut them some slack. Actually, they didn't need it: 87 percent of people over fifty could recall a relative's birthday, but less than 40 percent of people under thirty could do the same. A third of people in their twenties couldn't recall their own phone number and had to look it up on their digital device. And that was 2007, folks!

Then, after you forget the basic information that connects you to other human beings, Google takes care of the rest. A series of studies published in *Science* shows that when faced with a difficult memory task, instead of immediately reaching deep into our memories, our brains are now primed to think about computers—our first thought isn't "Gosh, I wonder if I can remember the name of that actor that's on the tip of my tongue," it's "Gosh, I should Google for the name of that actor on the tip of my tongue." And when you know there will be a way to find information later, you allow your memory to forget it. We've known this for years with lists—writing down your grocery list lets you forget what's on it. And Google functions as the great cultural grocery list. Because it lists everything, you can forget everything. The *Science* article likens technology-assisted remembering to a neuroprosthesis—the crutch of an artificial appendage that you can depend on to make up for a function that you have lost. Over time, wearing a back brace allows you to lose core strength, and over time, the support of Google can let your memory go equally flaccid.

The question is not whether technology has changed

and is continuing to change the human brain, but whether it's changing our brains in ways that are overall beneficial or detrimental. Do you really need to spend brain space remembering the names of U.S. presidents when you can look them up so simply? Might the brain space that in past years would have been used remembering state capitals and relatives' birthdays be repurposed for tasks like creativity and insight? And might the disruption of our sleep due to bright tablets be worth the sleep benefits of not having a partner on the other side of the bed polluting your sleep with light from a bedside lamp? Really, it's all too new to tell. If you can, put a peg in this question of technology's influence and let's revisit it in another twenty or fifty years . . . that is, if you can remember.

## *WHY YOU SHOULD STOP READING AND GO TO SLEEP*

Here you are again, snug in your bed. It's been quite a day—for your body, for your brain, and for the system of the two intertwined. Did you do good things for your brain today? Bad things? Did your brain do good things for you, helping you to make decisions, think thoughts, and take actions that make you proud in hindsight? Did you focus, resist temptation, solve problems, remember the things you needed to remember, play well with others, help to grow the brains of those around you, stay generally more happy than you were sad, or at least remain hopeful and optimistic even in the face of bad moods? Overall, do you feel that crisp crackling of your gray matter that is the mark of a day well spent?

Don't worry: you'll get another chance tomorrow.

And between now and then is a good night's sleep. But it seems kind of counterproductive: just think about how much more you could do if you stayed up! It isn't worth it, and here's why.

Back in the 1960s, when researchers were allowed to stick needles in people for thirty-eight nights and inject them with drugs like phenobarbital, a study from the Washington University School of Medicine showed that growth hormone secretion peaks during deep sleep. This hormone, so widely abused in professional sports, increases muscle mass and protein synthesis and stimulates the immune system (among other functions). If you want natural HGH, get a good night's sleep—it's one job of sleep to repair what you broke down during the day.

Sleep also allows the brain to clear itself of the chemical adenosine, which is a by-product of neuron functioning. While you think and act, adenosine accumulates, and while you sleep, it goes away. In fact, it's been shown that the accumulation of adenosine is one cue that makes the brain sleepy, and the invigorating effects of caffeine are due in part to the molecule's ability to block the effect of adenosine in the brain. The longer you stay up, the more you stew in your own adenosine, eventually turning your brain into sauerkraut.

And beauty sleep is for real too. Researchers in Stockholm took pictures of people after a normal night's sleep and again after thirty-one hours of wakefulness. Then they had sixty-five untrained people rate the attractiveness of these photos. Despite equivalent wardrobe and grooming, the pictures of sleep-deprived people were rated less healthy, less attractive, and more tired compared to the pictures of the same people after they'd slept.

The final benefit of sleep is something you've probably heard before: sleep helps you consolidate your memories. But recent research shows that memory consolidation during sleep goes beyond the idea of packaging memories in more efficient ways. Sleep may promote the brain's overall ability to be plastic—to adapt and transform itself to new stimuli—with more efficient memory being a by-product of this meta-level remodeling that happens overnight. You can see this in the brains of people learning to type: after practicing typing skills, people who slept for eight hours and then took a typing test did a heck of a lot better than people who did the same training and then stayed awake for eight hours.

From visual learning to emotional learning to traditional definitions of memory, sleep lets the brain change itself in ways that optimize what you need while letting go of what you don't. Basically, sleep is when the brain does the vast majority of the remodeling needed to make tomorrow different from today.

So turn off the light. Turn off the technology. Let go of the stress and machinations of the day—of all the things you should and shouldn't have done today, and all the things you should and shouldn't do tomorrow—and get a good night's sleep. Your brain will be better for it.

## Forgive Yourself to Fall Asleep

If you want to fall asleep, forgive yourself. When researchers looked at the thought control strategies of 406 college students, they found that withdrawing from unpleasant thoughts in favor of more pleasing ones works better than trying to control negative thoughts by beating yourself up over them. If your regrets or failures or worries or fears are keeping you awake right now, don't be ashamed—try to withdraw from these unpleasant thoughts and into gentler remembrances and hopes that can help you put worry to sleep along with your brain.